Lecture Notes in Mathematics

Edited by A. Dold and B. Eckmann

1120

Krzysztof Jarosz

Perturbations of Banach Algebras

Springer-Verlag
Berlin Heidelberg New York Tokyo

Author

Krzysztof Jarosz
Institute of Mathematics, Warsaw University
P.K.i.N. 9p., 00-901 Warsaw, Poland

AMS Subject Classification (1980): Primary: 46J05, 46J10, 46B20;
secondary: 46-02, 46E25, 46H05

ISBN 3-540-15218-0 Springer-Verlag Berlin Heidelberg New York Tokyo
ISBN 0-387-15218-0 Springer-Verlag New York Heidelberg Berlin Tokyo

Printing and binding: Beltz Offsetdruck, Hemsbach/Bergstr.
2146/3140-543210

Preface

This work is an introduction to a theory, in its initial stages; its development during the past few years may be seen in the works of Richard Rochberg, Barry E. Johnson as well as those of the author. We investigate various classes of small perturbations of algebraic structure of Banach algebras. We work with small isomorphisms between Banach algebras, with perturbations of multiplication as well as with other kinds of perturbations. In this paper we study the relations occuring between various types of perturbations and certain invariants of perturbations.

Much of the material presented here grew out of the author's conversations with Prof. Z. Sawoń. I am greatly indebted to him. Many thanks are also due to my wife Dorota for her careful reading of the manuscript.

Warsaw, January 1985

Krzysztof Jarosz

Table of Contents

PRELIMINARIES

§ 1. Introduction.

This is intended as an introduction to a theory that is starting to develop. We are concerned, roughly speaking, with the small deformations of the algebraic structure and "almost isometric" invariants in the theory of Banach algebras.

We consider three kinds of deformations: ε-perturbations of multiplication, ε-isomorphisms and ε-isometries.

By an <u>ε-perturbation</u> of a Banach algebra A we mean any multiplication $\times$ on the vector space A such that

$$\|f \times g - fg\| \leq \varepsilon \|f\| \cdot \|g\| \qquad \text{for all} \quad f,g \quad \text{in} \quad A.$$

By an <u>ε-isomorphism</u> between two Banach algebras A and B we mean any linear map T from A into B such that

$$\|T(fg) - T(f)T(g)\| \leq \varepsilon \|f\| \cdot \|g\| \qquad \text{for all} \quad f,g \quad \text{in} \quad A.$$

By an <u>ε-isometry</u> between two Banach spaces A and B we mean any linear continuous one to one map T from A onto B such that

$$\|T\| \leq 1 + \varepsilon \qquad \text{and} \qquad \|T^{-1}\| \leq 1 + \varepsilon.$$

Chapter I is devoted to the proof of the theorem stating that for any fixed function algebra A, ε-perturbations, ε-isomorphisms and ε-isometries produce the same class of algebras and hence we get a number of corollaries. In Chapter II we generalize some of the results of Chapter I to the theory of small into-isomorphisms. Chapter III is devoted to the study of small perturbations in semisimple, commutative Banach algebras. We define there a fairly large class of natural Banach algebras, we prove that in this class any isometry (of Banach spaces) is automatically an isomorphism (of Banach algebras) but this could not be generalized to ε-isometries. In Chapter IV we investigate the properties of Banach space X which ensure that every isometry (ε--isometry) from the algebra of all bounded linear operators on X onto itself is an algebra isomorphism (ε-isomorphism). In Chapter V we investigate algebraic properties of Banach algebras which are invariant under small perturbations.

The perturbation theory we present here has its source in old and classical theorems, but in fact its development started only in the last few years. As a first step in this direction one might consider the classical Banach-Stone theorem:

Theorem (Banach-Stone). Banach spaces $C(S)$ and $C(S')$ are isometric if and only if spaces S and S' are homeomorphic and hence if and only if they are isomorphic in the category of Banach algebras.

This theorem was generalized in three directions:

Theorem (Nagasawa 1959). Function algebras A and B are isometric if and only if they are isomorphic in the category of algebras.

Theorem (Amir, Cambern 1965). Assume that the Banach-Mazur distance between Banach spaces $C(S)$ and $C(S')$ is less than 2 then these spaces are actually isometric.

Theorem (Holsztyński 1966). Let S and S' be compact metric spaces, then any isometric embedding from $C(S)$ into $C(S')$ is induced by a continuous map from a compact subset of S' onto S.

On the other hand, independently, in the seventies, relatively common works appeared analizing, roughly speaking, the following "perturbation problems":

(i) Let A and B be Banach algebras, both contained in a Banach algebra C (e.g. A,B are C^*-algebras in $B(H)$). Assume A and B are close in a geometric sense (distance between unit balls of A and B is small). Does it follow that A and B are isomorphic as algebras or at least have some common algebraic properties?

(ii) Assume that we have defined two multiplications $\cdot$ and $\times$ on a Banach space A, which are close as bilinear maps. What common algebraic properties of $(A,\cdot)$ and $(A,\times)$ must appear?

In the paper we present a theory which allows a clear insight into all the generalizations of the Banach-Stone theorem as well as problems (i) and (ii) as various aspects of the same phenomenon (at least for function algebras). As a result of this the main emphasis was placed on the exploration of general relations occuring between various types of perturbations (Chapter I - IV). Specifically, the study of the perturbations of certain algebras as well as the study of certain perturbation invariants have been left untill Chapter V. In presenting Chapter V, we have taking adventage of the existing work of R. Rochberg [6] pertaining to perturbations of uniform algebras of analytic functions defined on Riemann surfaces, as well as the work of B.E. Johnson [3]; for these reasons only the most important facts pertaining to perturbations of algebras of analytic functions, often without proofs, have been presented. The main emphasis is placed on certain examples and perturbation invariants concerning the whole of function algebras.

The theory presented herein is only at the initial stages of developments; it is by no means complete and present more problem than solutions.

The paper is based on the author's most recent papers, it contains some unpublished results as well as some of the results of R. Rochberg and B. E. Johnson.

At the end of each chapter we give a historical and biographical note. We end the paper with a list of open problems.

In the paper we will make a free use of the elementary theory of analytic functions and quasiconformal maps to be found in L. Ahlfors [1], A. I. Markuszewicz [1] and of the theory of uniform algebras to be found in T. W. Gamelin [1], I. Suciu [1] and also of functional analysis to be found in W. Rudin [1].

I. PERTURBATIONS OF MULTIPLICATIONS AND ONTO-ISOMORPHISMS

§ 2. Basic facts.

2.1. Definition. Let A be a Banach algebra. By an ε-perturbation (algebraic ε-perturbation) of A we mean any multiplication × defined on the Banach space A such that

$$\|f \times g - fg\| \leq \varepsilon \|f\| \cdot \|g\| \qquad \text{for all } f,g \text{ in } A. \tag{1}$$

The ε-perturbations were studied by many authors (see e.g. Johnson [2], Jarosz [2-4,8-10], Rochberg [5-7]). The main problem in this field is, whether all multiplications on a Banach algebra A, near the given multiplication, share any of the properties of the original one. In this chapter we investigate a more general problem. We formulate a number of other natural definitions of small perturbations of function algebras, we find them equivalent and as corollaries we get theorems concerning the stability of the algebraic structure of a function algebra and theorems on uniqueness of multiplication in such algebras.

Let us start by formulating some general facts about algebraic ε-perturbations. Notice that if × is an ε-perturbation of a Banach algebra $(A,\cdot,\|\cdot\|)$ then $(A,\times,\|\cdot\|)$ need not be a Banach algebra (because the original norm on A need not be ×-submultiplicative) but the new multiplication is certainly continuous. It is well known that in this situation there exists an equivalent norm N on A which is ×-submultiplicative. If A is a uniform algebra this new norm may possess some special additional properties. By (1) we have

$$(1-\varepsilon)\|f\|^2 \leq \|f \times f\| \leq (1+\varepsilon)\|f\|^2 .$$

Hence, by induction, the norms $\|\cdot\|$ and N being equivalent, we can estimate the spectral radius of f in the algebra (A,×,N)

$$(1-\varepsilon)\|f\| \leq \text{spectral radius } (f) \leq (1+\varepsilon)\|f\|.$$

Hence by the theorem of Hirschfeld and Zelazko [1] the multiplication × is commutative for $\varepsilon < 1$ and the spectral radius is an equivalent, uniform norm on A.

Notice that we do not assume in the definition of an ε-perturbation the commutativity of the new multiplication but we get it from condition (1) for $\varepsilon < 1$.

We obtain the following proposition.

2.2. Proposition. Let $(A,\cdot,\|\cdot\|)$ be a uniform algebra and let $\times$ be an algebraic ε-perturbation of A with $\varepsilon < 1$. Then there exists an equivalent norm $|||\cdot|||$ on the Banach space A such that

$$(1-\varepsilon)\|f\| \leq |||f||| \leq (1+\varepsilon)\|f\|$$

for all f in A, and such that $(A,\times,|||\cdot|||)$ is a uniform algebra.

We will call the algebra $(A,\times,|||\cdot|||)$ an ε-deformation of A.

The algebra $(A,\times,|||\cdot|||)$ is a uniform algebra so it is isometrically isomorphic to a subalgebra B, of the space $C(S)$ for some compact Hausdorff space S.

Denote by T the identity operator from $(A,\cdot,\|\cdot\|)$ onto $(A,\times,|||\cdot|||)$ composed with an isometry of $(A,\times,|||\cdot|||)$ onto $(B,\cdot,\|\cdot\|)$. We have

$$T: A \to B, \qquad \|T\| \leq 1+\varepsilon, \qquad \|T^{-1}\| \leq \frac{1}{1-\varepsilon}$$

and

$$f \times g = T^{-1}(Tf \cdot Tg) \qquad \text{for all } f,g \text{ in } A.$$

So we have proven the following proposition.

2.3. Proposition. For any function algebra A and any ε-perturbation $\times$ of A, with $\varepsilon < 1$, this new multiplication is of the form

$$f \times g = T^{-1}(Tf \cdot Tg) \qquad \text{for all } f,g \text{ in } A$$

where T is a linear map between A and some function algebra B with $\|T\| \leq 1+\varepsilon$, $\|T^{-1}\| \leq \frac{1}{1-\varepsilon}$.

This suggests the following definition.

2.4. Definition. By a metric ε-perturbation of a function algebra A we mean any multiplication $\times$ defined on the Banach space A such that $\times$ is of the form

$$f \times g = T^{-1}(Tf \cdot Tg) \qquad \text{for all } f,g \text{ in } A$$

where T is a linear map between A and some function algebra B (called a metric ε-deformation of A) such that $\|T\| \leq 1+\varepsilon$ and $\|T^{-1}\| \leq 1+\varepsilon$.

•

Proposition 2.3 states that any small algebraic perturbation of a function algebra is also a small metric perturbation of this algebra. The following example shows that the converse implication does not hold.

2.5. Example. Let A be a function algebra such that there is a $\chi \in A^{-1}$ with $\|\chi\| = 1$, $\|\chi^{-1}\| = 1$ and $\|\chi - 1\| \geq 1$. The map $T: A \to A$: $Tf = \chi \cdot f$ is an onto isometry but the new multiplication on A defined by T is of the form $f \times g = T^{-1}(Tf \cdot Tg) = \chi \cdot f \cdot g$ so it is not an algebraic ε-perturbation of the original one (if $\varepsilon < 1$).

•

To avoid this trivial non equivalence we have two possibilities. The first one is to add to Definition 2.4 the following condition

$$\|1_B - T1_A\| \leq \varepsilon, \tag{*}$$

and the second one is to change the first definition to the effect:

2.6. Definition. Let A be a Banach algebra and let $\times$ be another multiplication on the Banach space A. We call $\times$ an ε-perturbation of A if there is a $\chi \in A^{-1}$ such that $\|\chi\| \leq 1+\varepsilon$, $\|\chi^{-1}\| \leq 1+\varepsilon$ and

$$\|\chi \cdot f \cdot g - f \times g\| \leq \varepsilon \|f\| \cdot \|g\| \qquad \text{for any } f,g \text{ in } A.$$

•

Before we establish a relation between the above definitions let us first look for "reasonable" weaker definitions of small perturbations of multiplication.

2.7. Definition. Let A be a Banach algebra and let $\times$ be any multiplication on the Banach space A. We call $\times$ an ε-perturbation of A if

$$\big|\,\|f \times g\| - \|f \cdot g\|\,\big| \leq \varepsilon\|f\|\cdot\|g\| \qquad \text{for any } f,g \text{ in } A.$$

•

Notice please that using the same method as in the proof of Proposition 2.2 it is easy to show that for any function algebra A and any its small perturbation defined as above, the algebra $(A,\times)$ is commutative and the spectral norm is equivalent to the original one, but $(A,\times)$ is not necessary a function algebra because it may not to have a unit. (To get a simple example let $A = C(S)$ for some non one point compact set S. Let $\chi \in A \setminus A^{-1}$ be such that $|\chi| \equiv 1$ on S and put

$$f \times g = \chi^2 \cdot f \cdot g \qquad \text{for } f,g \text{ in } A.)$$

2.8. Definition. Let A be a Banach algebra with unit $\mathbf{1}$ and let $\times$ be any multiplication on the Banach space A with unit e. We call $\times$ an ε-perturbation of A if

$$\|f \times g\| \leq (1 + \varepsilon)\|f\|\cdot\|g\| \qquad \text{for all } f,g \text{ in } A$$

and

$$\|\mathbf{1} - e\| \leq \varepsilon.$$

•

The main theorem of this chapter states that though we have six definitions of ε-perturbations, there are, in fact, only two distinct kinds of perturbations: the metric and the algebraic; and even these two kinds are very close to each other.

§ 3. The main theorem.

The following theorem establish relations between the various definitions of small perturbations.

3.1. THEOREM. Let A be a Banach algebra with unit $\mathbb{1}$ and let $\times$ be any multiplication on A with the unit denoted by e. Then the following conditions are equivalent:

(i) $\big|\,\|f \times g\| - \|f \cdot g\|\,\big| \le \varepsilon_1 \|f\| \cdot \|g\|$ for all f,g in A;

(ii) there is a $\chi \in A^{-1}$ with $\|\chi\| \le 1 + \varepsilon_2$, $\|\chi^{-1}\| \le 1 + \varepsilon_2$, such that

$\|\chi \cdot f \cdot g - f \times g\| \le \varepsilon_2 \|f\| \cdot \|g\|$ for all f,g in A.

If A is a function algebra then the above conditions are also equivalent to:

(iii) there is a linear continuous map T between A and some function algebra B such that $\|T\| \le 1 + \varepsilon_3$, $\|T^{-1}\| \le 1 + \varepsilon_3$ and $\times$ is of the form

$f \times g = T^{-1}(Tf \cdot Tg)$ for all f,g in A.

If A is a function algebra and $\times$ is such that

(e) $\|\mathbb{1} - e\| \le \varepsilon_3$

then (iii) is also equivalent to

(iv) $\|f \times g - f \cdot g\| \le \varepsilon_4 \|f\| \cdot \|g\|$ for f,g in A.

If A is a complex function algebra and (e) is satisfied then (iii) is also equivalent to

(v) $\|f \times g\| \le (1 + \varepsilon_5) \|f\| \cdot \|g\|$ for all f,g in A.

Moreover if A is a function algebra and conditions (e) and (iii) are satisfied then there is a homeomorphism φ between the Choquet boundaries of the algebras B and A such that

(φ) $|Tf(s) - f \circ \varphi(s)| \le \varepsilon_6 \|f\|$ for all $f \in A$ and $s \in \mathrm{Ch}B$.

Where ε_i: $i = 1, \ldots, 6$ tend to zero simultaneously and ε_i are not greater than some absolute constant ε_0.

Let us recapitulate what the above theorem says:

1. For any Banach algebra with a unit and any new multiplication defined on the underlying Banach space which possesses a unit Definitions 2.6 and 2.7 are equivalent i.e. ε from Definition 2.6 and ε from Definition 2.7 tend to zero simultaneously.

2. For any function algebra Definitions 2.4 (metric perturbation) and 2.6 are equivalent.

3. For any complex function algebra Definitions 2.1 (algebraic perturbation) and 2.8 are equivalent.

4. For any complex function algebra and its small perturbation which satisfies (*) or equivalently (e) Definitions 2.1, 2.4, 2.6, 2.7, and 2.8 are equivalent.

5. Any small perturbation of a function algebra is given by an "almost composition" with a homeomorphism of the Choquet boundaries.

Assuming that the new multiplications possess the same unit that the original one, for complex function algebras, we can formulate the above theorem also in the following way.

<u>3.1'. THEOREM</u>. Let A be a complex function algebra and let $\times_n$, $n \in \mathbb{N}$ be a sequence of multiplications defined on a Banach space A. Assume that all the multiplications $\times_n$, $n \in \mathbb{N}$ and the original one possess the same unit. The following are equivalent:

(i) there is a sequence of positive numbers $(\varepsilon_n)_{n=1}^{\infty}$ with $\lim_{n\to\infty} \varepsilon_n = 0$ and such that

$$\big|\, \|f \times_n g\| - \|f \cdot g\| \,\big| \le \varepsilon_n \|f\| \cdot \|g\| \qquad \text{for all } n \in \mathbb{N}, \quad f,g \in A;$$

(ii) there is a sequence of positive numbers $(\varepsilon_n)_{n=1}^{\infty}$ with $\lim_{n\to\infty} \varepsilon_n = 0$ and such that

$$\|f \times_n g - f \cdot g\| \le \varepsilon_n \|f\| \cdot \|g\| \qquad \text{for all } n \in \mathbb{N}, \quad f,g \in A;$$

(iii) there is a sequence of positive numbers $(\varepsilon_n)_{n=1}^{\infty}$ with $\lim_{n\to\infty} \varepsilon_n = 0$ and such that

$$\|f \times_n g\| \leq (1+\varepsilon_n)\|f\|\cdot\|g\| \quad \text{for all} \quad n \in \mathbb{N}, \quad f,g \in A;$$

(iv) there is a sequence $(B_n)_{n=1}^{\infty}$ of function algebras and a sequence $(T_n)_{n=1}^{\infty}$ of linear onto isomorphisms from A onto B_n, respectively, such that

$$\lim_n \|T_n\|\,\|T_n^{-1}\| = 1 \quad \text{and}$$

$$f \times_n g = T_n^{-1}(T_n f \cdot T_n g) \quad \text{for all} \quad n \in \mathbb{N}, \quad f,g \in A.$$

Furthermore if T_n, B_n are as in (iv) then there is a sequence $(\varphi_n)_{n=1}^{\infty}$ of homeomorphisms between Choquet boundaries of B_n and A, respectively, and a sequence of positive numbers $(\varepsilon_n)_{n=1}^{\infty}$ with $\lim_{n\to\infty} \varepsilon_n = 0$ and such that

$$|T_n f(s) - f \circ \varphi_n(s)| \leq \varepsilon_n \|f\| \quad \text{for all} \quad n \in \mathbb{N},\ f \in A, \ \text{and} \quad s \in \mathrm{Ch}B_n.$$

The proof of Theorem is rather long so in this section we shall deduce some corollaries and give examples defering the proof to §4.

The first corollary states that the multiplication on a function algebra is uniquely determined by its geometry.

3.2. Corollary. Suppose $(A, \cdot, \|\cdot\|)$ is a complex function algebra and let $\times$ be any multiplication on the Banach space A with the same unit and such that $(A, \times, \|\cdot\|)$ is a Banach algebra (this means such that $\|f \times g\| \leq \|f\|\cdot\|g\|$ for all f,g in A), then the new multiplication $\times$ and the original one coincide.

Proof. By Theorem 3.1 (iv) ⇔ (v).

The above corollary can also be formulated in the following way, giving a generalization of the Nagasawa Theorem (Nagasawa [1]) stating that any linear isometry between two function algebras preserving units is also an isomorphism in the category of Banach algebras.

3.3. Corollary. Let A be any complex function algebra with unit $\mathbb{1}_A$ and let B be any Banach algebra with unit $\mathbb{1}_B$. Suppose T is a linear isometry from A onto B such that $T\mathbb{1}_A = \mathbb{1}_B$. Then T is an algebra isomorphism.

Notice that in the above corollaries we have considered only the complex algebras. The following example shows that they are not valid for the real function algebras even in two dimensions.

3.4. Example. Let $A = (\mathbb{R}^2, \cdot, \|\cdot\|_\infty)$ be the two dimensional real function algebra and define a bilinear map

$$\rho_t \colon \mathbb{R}^2 \times \mathbb{R}^2 \to \mathbb{R}^2 \quad \text{by}$$

$$\begin{aligned}\rho_t((x,y),(x',y')) = (&x \cdot x' - t \cdot (x-y) \cdot (x'-y'),\\ &y \cdot y' - t \cdot (x-y) \cdot (x'-y')).\end{aligned}$$

A direct computation shows that for $0 \leq t \leq \frac{1}{2}$ the bilinear map ρ_t is a commutative, associative multiplication on $\mathbb{R}^2$ such that $\|\rho_t\| = 1$ and

$$\rho_t((1,1),(x,y)) = (x,y) \qquad \text{for any} \qquad (x,y) \in \mathbb{R}^2.$$

•

Corollary 3.3 shows that in the Nagasawa theorem it is not necessary to assume that both algebras A and B are uniform algebras but it is sufficient to assume that only one of them is uniform and the second is an arbitrary Banach algebra with unit (we do not even assume commutativity).

Theorem 3.1 also provides a generalization of the Nagasawa Theorem of another type showing that this theorem is in some sense stable.

For this purpose we need two definitions.

3.5. Definition. We shall call two Banach spaces A and B ε-isometric if there is a linear continuous map $T \colon A \to B$ with $\|T\| \cdot \|T^{-1}\| \leq 1 + \varepsilon$. We will call such a T an ε-isometry.

•

3.6. Definition. We shall call two Banach algebras A and B ε-almost algebraically isomorphic if there is a linear map $T \colon A \to B$

such that

$$\|T^{-1}(Tf \cdot Tg) - f \cdot g\| \leq \varepsilon \|f\| \cdot \|g\| \quad \text{for all } f,g \text{ in } A.$$

We will call such a T an ε-almost algebra isomorphism. •

3.7. Corollary. There is a positive constant ε_0 such that for any function algebras A and B we have

a) if there is an ε-isometry T from A onto B then there is an ε'-isometry $\tilde{T}$ from A onto B such that $\tilde{T}(\mathbf{1}) = \mathbf{1}$;

b) if $\tilde{T}: A \to B$ is an onto ε'-isometry such that $\tilde{T}(\mathbf{1}) = \mathbf{1}$ then $\tilde{T}$ is an ε''-almost algebra isomorphism;

c) if $S: A \to B$ is an ε''-almost algebra isomorphism with $\varepsilon'' < \varepsilon_0$ then it is an ε'''-isometry.

Here $\varepsilon, \varepsilon', \varepsilon''$, and ε''' tend to zero simultaneously.

Proof. We prove first a) and b) together. Assume there is a $T: A \to B$ with $\|T\| \cdot \|T^{-1}\| \leq 1 + \varepsilon$. Without loss of generality we can assume that $\|T\| \leq 1 + \varepsilon$ and $\|T^{-1}\| \leq 1 + \varepsilon$ (if T preserves units then we get it automatically from the inequality $\|T\| \cdot \|T^{-1}\| \leq 1 + \varepsilon$). By Theorem 3.1 ((ii) $\Leftrightarrow$ (iii)) there is a $\chi \in A^{-1}$ with $\|\chi\| \leq 1 + \delta$, $\|\chi^{-1}\| \leq 1 + \delta$ and such that

$$\|\chi \cdot f \cdot g - T^{-1}(Tf \cdot Tg)\| \leq O(\varepsilon) \cdot \|f\| \cdot \|g\| \quad \text{for all } f,g \text{ in } A \tag{2}$$

where $\delta = O(\varepsilon)$.

Put $\tilde{T}: A \to B$: $\tilde{T}f = T(f \cdot \chi^{-1})$ and notice that if $T\mathbf{1}_A = \mathbf{1}_B$ then by Theorem 3.1 ((iii) $\Leftrightarrow$ (iv)) we can put $\chi = \mathbf{1}$, so then $\tilde{T} = T$. We have $\tilde{T}\mathbf{1} = \mathbf{1}$ and $\|\tilde{T}\| \cdot \|\tilde{T}^{-1}\| \leq (1 + \delta)^2 (1 + \varepsilon)$. From (2) we get

$$\|f \cdot g - \tilde{T}^{-1}(\tilde{T}f \cdot \tilde{T}g)\| = \|f \cdot g - \chi \cdot T^{-1}(T(f \cdot \chi^{-1}) \cdot T(g \cdot \chi^{-1})\| \leq$$

$$\leq \|\chi\| \cdot \|\chi \cdot (f \cdot \chi^{-1}) \cdot (g \cdot \chi^{-1}) - T^{-1}(T(f \cdot \chi^{-1}) \cdot T(g \cdot \chi^{-1})\| \leq$$

$$\leq (1 + \delta)\delta \|f \cdot \chi^{-1}\| \cdot \|g \cdot \chi^{-1}\| \leq (1 + \delta)^3 \|f\| \cdot \|g\| = \varepsilon' \|f\| \cdot \|g\|.$$

So we have proven that $\tilde{T}$ is an ε'-almost algebra isomorphism.

To prove c) assume now that there is a linear map $S: A \to B$ such that

$$\|f \times g - f \cdot g\| \leq \varepsilon \|f\| \cdot \|g\| \qquad \text{for all } f,g \text{ in } A \tag{3}$$

where $f \times g = S^{-1}(Sf \cdot Sg)$.

Put $e = S^{-1}(\mathbb{1}_B)$, it is the unit of the algebra $(A, \times)$. From (3) for $f = \mathbb{1}$ and $g = e$ we get

$$\|e\| - 1 \leq \|e - \mathbb{1}\| \leq \varepsilon \|e\|$$

hence

$$\|e\| \leq 1/(1-\varepsilon)$$

and finally

$$\|e - \mathbb{1}\| \leq \varepsilon/(1-\varepsilon).$$

So by Theorem 3.1 ((iv) $\Leftrightarrow$ (iii)) there is a continuous linear map T between A and some function algebra C such that $f \times g = T^{-1}(Tf \cdot Tg)$ and $\|T\| \leq 1 + \eta$, $\|T^{-1}\| \leq 1 + \eta$ with $\eta = O(\varepsilon)$. Maps $T \circ S^{-1}$ and $S \circ T^{-1}$ are algebra isomorphisms of two function algebras B and C so they are isometries and hence

$$\|S\| \cdot \|S^{-1}\| = \|T^{-1}\| \cdot \|T\| \leq (1+\eta)^2 = O(\varepsilon)$$

and this means that S is an ε'-almost isometry from A onto B.

•

<u>3.8. Remark</u>. We do not assume the continuity of the map S but we get it from the condition

$$\|S^{-1}(Sf \cdot Sg) - f \cdot g\| \leq \varepsilon \|f\| \cdot \|g\| \qquad \text{for all } f,g \text{ in } A$$

for sufficiently small ε (in fact for $\varepsilon < 1$).

•

We can also formulate Corollary 3.7 in the following way using the definitions of metric and algebraic perturbations.

<u>3.9. Corollary</u>. For any function algebra A the small metric and algebraic perturbations of A produce the same class of algebras.

Corollary 3.8 does not settle the problem only in a limit situation. Namely it does not show whether the Banach-Mazur distance between two function algebras is equal to one if and only if these algebras are algebraically isomorphic. This "limit question" was first posed by R. Rochberg [5]. He defined and investigated a natural measure of closeness of two algebras A and B:

$$D(A,B) = \inf\{\varepsilon: \ B \text{ is algebraically isomorphic to an algebraic } \varepsilon\text{-perturbation of } A\}.$$

So our question reads: Does $D(A,B) = 0$ implies A and B are algebraically isomorphic?

The answer is negative.

<u>3.10. Example</u>. Let us denote by $A(R,r)$ the algebra of functions continuous on the set $P(R,r)$ and analytic on $\operatorname{int} P(R,r)$, where

$$P(R,r) = \{z \in \mathbb{C}: \ r \le |z| \le R\} .$$

$A(R,r)$ is equipped with the uniform norm. Let $A_n = A(2+\frac{1}{n},1)$ for $n \in \mathbb{N}$, let $\tilde{A}_n$ be a direct sum of infinitely many copies of A_n and let

$$A = A_1 \oplus A_2 \oplus A_3 \oplus \dots$$
$$B = A \oplus A(2,1)$$

where the direct sum is taken in the sense of "convergent sequences" this means

$$\bigoplus_{n=1}^{\infty} C_n = \{(h_n)_{n=1}^{\infty}: \ h_n \in C_n, \quad \exists \lambda \in \mathbb{C} \quad \lim \|h_n - \lambda \mathbf{1}\| = 0\}$$

with a usual supremum norm.

We will prove that $D(A,B) = 0$ but A and B are not isometric.

If the algebras A and B were isometrically isomorphic then there would be a one to one correspondence between factors of A and the isomorphic factors of B but it is well known that the algebras $A(R,r)$ and $A(R',r')$ are isomorphic if and only if the sets $P(R,r)$ and $P(R',r')$ are conformly equivalent and this holds if and only if $R/r = R'/r'$, but there is no factor in A isomorphic to $A(2,1)$.

To prove $D(A,B) = 0$ it is sufficient, by Corollary 3.7, to prove that the Banach-Mazur distance between A and B is one. We define a sequence of operators T_k from A onto B.

$$T_k|_{\tilde{A}_n} = Id_{\tilde{A}_n} \qquad \text{for} \quad n \neq k$$

and

$$T_k|_{\tilde{A}_k} : \tilde{A}_k \to \tilde{A}_k \oplus A(2,1):$$

$$T_k|_{\tilde{A}_k}((f_1,f_2,\ldots)) = ((f_2,f_3,\ldots), S_k(f_1)),$$

where

$$S_k: A_k \to A(2,1) \qquad \text{is defined by}$$

$$S_k\left(\sum_{p=-\infty}^{+\infty} a_p z^p\right) = \sum_{p=-\infty}^{0} a_p z^p + \sum_{p=1}^{\infty} \left(1+\frac{1}{2k}\right)^p a_p z^p .$$

Now the equality $\lim \|T_k\|\cdot\|T_k^{-1}\| = 1$ is an immediate consequence of the well-known fact that $\|S_k\|\cdot\|S_k^{-1}\|$ tends to 1 with k tending to infinity.

Notice at the end of the example that the algebras A and B are very "nice". They are separable, their Choquet and Shilov boundaries coincide and both are generated by rational functions of one element.

●

<u>3.11. Corollary</u>. Let A be a Banach algebra with a unit. Assume that on the underlying Banach space A another multiplication $\times$ is defined, such that $\|f \times g\| = \|f \cdot g\|$ for any f,g in A. Then if the both multiplications $\times$ and $\cdot$ possess the same units then they coincide.

Proof. By Theorem 3.1 ((i) $\Leftrightarrow$ (iv)).

•

3.12. Corollary. If the Shilov and the Choquet boundaries of a uniform algebra A coincide then, for any sufficiently small perturbation $\times$ of A, the Choquet and the Shilov boundaries of $A_\times$ also coincide.

Proof. By Theorem 3.1 there is a homeomorphism φ between ChA and $ChA_\times$ so if ChA is compact then also $ChA_\times$ is compact.

•

3.13. Definition. Let A be a Banach algebra. We call A stable or rigid if there is an $\varepsilon_0 > 0$ such that for any ε_0-perturbation $\times$ of A algebras A and $(A,\times)$ are isomorphic.

•

3.14. Corollary. For any compact, Hausdorff space S the algebra C(S) is stable.

Proof. Let $\times$ be an ε-perturbation of C(S). By Theorem 3.1, if ε is sufficiently small, there are a linear isomorphism T from C(S) onto a closed subalgebra B of C(S) and a homeomorphism φ from S onto itself such that

$$T: (C(S), \times) \to B \quad \text{is an algebra isomorphism}$$

and

$$\|T(f) - f\circ\varphi\| \leq O(\varepsilon)\cdot\|f\| \quad \text{for any } f \in C(S).$$

The map

$$\Phi: C(S) \to C(S) : f \mapsto f\circ\varphi$$

is an onto isometry and we have $\|\Phi - T\| \leq O(\varepsilon)$. Hence, if ε is sufficiently small, we get T is onto C(S), so $B = C(S)$.

•

At the end of this part we give an example showing that the assertion of Theorem 3.1 cannot be strengthened to the effect "... there is a homeomorphism φ between the Shilov boundaries of the algebras B and A ...". This means that for any $\varepsilon > 0$ there are function algebras A and A_ε such that $D(A,A_\varepsilon) < \varepsilon$ but ∂A and ∂A_ε are not homeomorphic.

3.15. Example. Put $G = \{z \in \mathbb{C}: |z| \leq 1\} \cup \{z \in \mathbb{C}: \operatorname{Im} z = 0,\ 2 \leq |z| \leq 3\}$ and let A be the algebra of all continuous functions on G which are holomorphic on int G. For any $1 \geq \varepsilon \geq 0$ we put $A_\varepsilon = \{f \in A: f(0) = f(2),\ f(\varepsilon) = f(-2)\}$ and we define a linear operator $T_\varepsilon: A_\varepsilon \to A_0$ by

$$T_\varepsilon f(z) = \begin{cases} f(z) & \text{for } z \in G,\ \operatorname{Re} z > -2 \\ f(z) - f(-2) + f(0) & \text{for } z \in G,\ \operatorname{Re} z \leq -2. \end{cases}$$

It is easy to verify that $\|T_\varepsilon\| \cdot \|T_\varepsilon^{-1}\| \xrightarrow[\varepsilon \to 0]{} 1$ but ∂A_ε and ∂A_0 are not homeomorphic for $\varepsilon > 0$.

•

3.16. Remark. Using the above example and by the method of Example 3.10 it can be proved that there are two function algebras A and B such that $D(A,B) = 0$ but ∂A and ∂B are not homeomorphic.

•

§ 4. Proof of the main theorem.

Let $A, \cdot, \mathbf{1}, \times, e$ be as in Theorem 3.1 and denote by (F) the following condition

$$\left.\begin{array}{l} \text{For any } F \text{ in } \operatorname{Ch}A \text{ there is an } \tilde{F} \text{ in } \mathfrak{M}(A_\times) \text{ such that} \\ \|F - \tilde{F}\| \leq \varepsilon. \end{array}\right\} \quad (F)$$

We divide the proof of Theorem 3.1 into the following steps:

1. (v) + (e) ⇒ (F) — for complex function algebras;
2. (F) ⇒ (iv) — for function algebras;
3. (iv) ⇒ (iii) + (e) — for function algebras;
4. (iii) ⇒ (i) — for function algebras;

5. (i) $\Rightarrow$ (ii) for Banach algebras;
6. (ii) $\Rightarrow$ (i) $\Rightarrow$ (v) for Banach algebras;
7. (ii) + (e) $\Rightarrow$ (iv) for Banach algebras;
8. (iv) $\Rightarrow$ (F) for real function algebras;
9. (F) $\Rightarrow$ (φ) for function algebras.

At various points of the proof we use inequalities involving ε which are only valid if ε is sufficiently small, in these circumstances we will merely assume that ε is near 0 and this assumption gives rise to the constant ε_o.

<u>Step 1.</u> <u>(v) + (e) $\Rightarrow$ (F) for complex function algebras.</u>

<u>Proof</u>. We first assume that $e = \mathbb{1}$ and

$$\|f \times g\| \leq (1+\varepsilon)\|f\|\cdot\|g\| \quad \text{for all } f,g \text{ in } A.$$

Let us introduce some notation

$$\Omega = \{x+iy \in \mathbb{C}: (x-\tfrac{1}{2})^2 + (y-\tfrac{1}{2})^2 < \tfrac{1}{2}\} \cap$$

$$\cap \{x+iy \in \mathbb{C}: (x-\tfrac{1}{2})^2 + (y+\tfrac{1}{2})^2 < \tfrac{1}{2}\}$$

and for $r > 0$

$$D(r) = \{x+iy \in \mathbb{C}: x^2+y^2 < r^2\}.$$

Notice that without loss of generality we may assume that A is an algebra of continuous functions on ∂A.

Fix $\delta > 0$, and let $\kappa: \overline{D(1)} \to \overline{\Omega}$ be a continuous map of $\overline{D(1)}$ onto $\overline{\Omega}$ such that κ is analytic on $D(1)$ and

$$\kappa(1) = 1 \quad \text{and} \quad \kappa(0) = \frac{\delta}{2}.$$

Let $V \subset \mathbb{C}$ be a neighbourhood of 0 such that

$$\kappa(V) \subset \overline{\Omega} \cap D(\delta).$$

Now fix any point $s_o \in \mathrm{Ch}A$ and any of its neighbourhoods $U \subset \partial A$, and let $f \in A$ be such that

$$\|f\| = f(s_o) = 1 \quad \text{and} \quad f(\partial A \setminus U) \subset V.$$

The function $\kappa \circ f \in A$ has the following properties:

a) $\kappa \circ f(\partial A) \subset \overline{\Omega}$;

b) $\|\kappa \circ f\| = \kappa \circ f(s_o) = 1$;

c) $\kappa \circ f(\partial A \setminus U) \subset \overline{\Omega} \cap D(\delta)$.

Hence for any $s_o \in ChA$ there exists a net $(f_\alpha) \subset A$ such that

A) $f_\alpha(\partial A) \subset \overline{\Omega}$,

B) $\|f_\alpha\| = f_\alpha(s_o) = 1$,

C) (f_α) tends uniformly to zero on the compact subsets of the set $\partial A \setminus \{s_o\}$.

Using the net (f_α) we define

$$g'_\alpha = f_\alpha + i(\mathbb{1} - f_\alpha) ,$$

$$g''_\alpha = f_\alpha - i(\mathbb{1} - f_\alpha).$$

By a direct computation

$$g'_\alpha \times g'_\beta = f_\alpha + f_\beta - \mathbb{1} + i(f_\alpha + f_\beta - 2f_\alpha \times f_\beta).$$

Further observe that, by the definition of Ω, we have

$$\|g'_\alpha\| = \sup_{s \in \partial A} |f_\alpha(s) + i(1 - f_\alpha(s))| \leq \sup_{z \in \Omega} |z + i(1 - z)| = 1.$$

Hence from (v) we get

$$\begin{aligned} 1 + \varepsilon &\geq \|g'_\alpha \times g'_\beta\| \geq |g'_\alpha \times g'_\beta(s_o)| = \\ &= |1 + 2i(1 - f_\alpha \times f_\beta(s_o))|. \end{aligned} \tag{4}$$

The same computations for the functions g''_α and g''_β show that

$$\begin{aligned} 1 + \varepsilon &\geq \|g''_\alpha \times g''_\beta\| \geq |g''_\alpha \times g''_\beta(s_o)| = \\ &= |1 - 2i(1 - f_\alpha \times f_\beta(s_o))| . \end{aligned} \tag{5}$$

Inequalities (4) and (5) can be satisfied simultaneously only if

$$|1 - f_\alpha \times f_\beta(s_o)| \leq \sqrt{\varepsilon/2 + \varepsilon^2/4} \leq \sqrt{\varepsilon}\,. \tag{6}$$

Now, for any $g \in A$ define two functionals $T^l_g: A \to \mathbb{C}$ and $T^r_g: A \to \mathbb{C}$

$$T^l_g(f) = g \times f(s_o), \qquad T^r_g(f) = f \times g(s_o).$$

For each $g \in A$ fix two regular measures μ^l_g and μ^r_g on ∂A such that

$$\mu^l_g(f) = T^l_g(f), \qquad \mathrm{var}\,(\mu^l_g) = \|T^l_g\|,$$

$$\mu^r_g(f) = T^r_g(f), \qquad \mathrm{var}\,(\mu^r_g) = \|T^r_g\| \qquad \text{for all } f \text{ in } A.$$

Inequality (6) shows that

$$|\mu^l_{f_\alpha}(f_\beta) - 1| \leq \sqrt{\varepsilon} \qquad \text{for any } \alpha \text{ and all } \beta. \tag{7}$$

By the definition of (f_α) we get

$$|\mu^l_{f_\alpha}(\{s_o\}) - 1| \leq \sqrt{\varepsilon}.$$

Hence, because $\mathrm{var}\,(\mu^l_{f_\alpha}) = \|T^l_{f_\alpha}\| \leq 1 + \varepsilon$, the measure $\mu^l_{f_\alpha}$ is of the form

$$\mu^l_{f_\alpha} = \delta_{s_o} + \Delta\mu^l_{f_\alpha} \tag{8}$$

where δ_{s_o} is a Dirac measure concentrated at the point s_o and $\mathrm{var}\,(\Delta\mu^l_{f_\alpha}) \leq 3\sqrt{\varepsilon}$.

Now let g_o be any element of A such that $\|g_o\| = 1 = g_o(s_o)$. By (8) we get

$$\mu^r_{g_o}(f_\alpha) = f_\alpha \times g_o(s_o) = \mu^l_{f_\alpha}(g_o) = g_o(s_o) + \Delta\mu^l_{f_\alpha}(g_o) = 1 + \Delta\mu^l_{f_\alpha}(g_o)$$

and hence $|\mu^r_g(f_\alpha) - 1| \leq 3\sqrt{\varepsilon}$.

In the same way as before we get

$$\mu^r_{g_o} = \delta_{s_o} + \Delta\mu^r_{g_o} \qquad \text{where} \qquad \mathrm{var}\,(\Delta\mu^r_{g_o}) \leq 7\sqrt{\varepsilon}\,. \tag{9}$$

Using this we can estimate the norm of $g_o \times g_o$ from below

$$\|g_o \times g_o\| \geq |g_o \times g_o(s_o)| = |\mu^r_{g_o}(g_o)| = |1 + \Delta\mu^r_{g_o}(g_o)| \geq 1 - 7\sqrt{\varepsilon} .$$

Because s_o is an arbitrary point of ChA this proves that

$$\|g \times g\| \geq (1 - 7\sqrt{\varepsilon})\|g\|^2 \qquad \text{for any } g \text{ in } A. \tag{10}$$

As an immediate consequence of (10) we conclude that the spectral radius of any element g of the algebra $(A,\times)$ is not less than $(1 - 7\sqrt{\varepsilon})\|g\|$. Hence by the theorem of Hirschfeld and Żelazko [1] one obtains commutativity of the multiplication $\times$ (if $1 - 7\sqrt{\varepsilon} > 0$).

Applying (10) for $g = f_\alpha$ and using the commutativity of $\times$ we find that there exists a linear and $\times$-multiplicative functional $\tilde{F}_\alpha$ on A such that $|\tilde{F}_\alpha(f_\alpha)| \geq 1 - 7\sqrt{\varepsilon}$. For any f in A of norm one we have

$$(1 + \varepsilon)\|\tilde{F}_\alpha\| \geq \|\tilde{F}_\alpha\| \cdot \|f \times f\| \geq |\tilde{F}_\alpha(f \times f)| = |\tilde{F}_\alpha(f)|^2 ,$$

hence

$$(1 + \varepsilon)\|\tilde{F}_\alpha\| \geq \|\tilde{F}_\alpha\|^2, \quad \text{and so} \quad \|\tilde{F}_\alpha\| \leq 1 + \varepsilon.$$

Let ν_α be a regular measure on ∂A which represents the functional $\tilde{F}_\alpha$ and such that $\mathrm{var}\,(\nu_\alpha) = \|\tilde{F}_\alpha\|$. We have

$$\left.\begin{aligned} &|\nu_\alpha(f_\alpha)| \geq 1 - 7\sqrt{\varepsilon} , \\ &\mathrm{var}\,(\nu_\alpha) \leq 1 + \varepsilon , \\ &\nu_\alpha(\mathbb{1}) = 1 \qquad\qquad \text{for all indices } \alpha. \end{aligned}\right\} \tag{11}$$

Taking a net finer than (f_α) and using the weak $*$ compactness of ∂A we can assume, without loss of generality, that the net (ν_α) is weak $*$ convergent to the measure ν_o. The measure ν_o also represents a linear and $\times$-multiplicative functional $\tilde{F}_o$ on A. From (11) we derive that the measure ν_o is of the form

$$\nu_o = \delta_{s_o} + \Delta\mu_{s_o} \qquad \text{where} \qquad \mathrm{var}\,(\Delta\mu_{s_o}) \leq c_1\sqrt{\varepsilon} = \varepsilon' .$$

Since s_o is an arbitrary point of ChA the above statement proves (F) in the case $\mathbb{1} = e$.

To end the proof of Step 1 let $F \in$ ChA and let $\times$ be any multiplication on A such that

$$\|f \times g\| \leq (1 + \varepsilon)\|f\|\cdot\|g\| \qquad \text{for all } f,g \text{ in } A$$

and

$$\|\mathbb{1} - e\| \leq \varepsilon.$$

If ε is sufficiently small $(\varepsilon(1 + \varepsilon) < 1)$ then the above implies that $\mathbb{1}$ is an invertible element of the algebra $(A,\times)$ and

$$\mathbb{1}^{-1} = e + (e - \mathbb{1}) + (e - \mathbb{1}) \times (e - \mathbb{1}) + \ldots .$$

Hence

$$\|\mathbb{1}^{-1}\| \leq \|e\| + \sum_{n=1}^{\infty} \|(e - \mathbb{1}) \times \ldots \times (e - \mathbb{1})\| \leq$$
$$\leq 1 + \varepsilon + \sum_{n=1}^{\infty} \varepsilon^n (1 + \varepsilon)^{n-1} = 1 + \varepsilon + \frac{\varepsilon}{(1 - \varepsilon(\varepsilon + 1))} . \tag{12}$$

We define a new multiplication $\hat{\times}$ on A by

$$f \hat{\times} g = \mathbb{1}^{-1} \times f \times g \qquad \text{for } f,g \text{ in } A.$$

The multiplication $\hat{\times}$ has the same unit as the original multiplication of the function algebra A and by (12) we have

$$\|f \hat{\times} g\| \leq (1 + \varepsilon)^2 \|\mathbb{1}^{-1}\|\cdot\|f\|\cdot\|g\| \leq (1 + O(\varepsilon))\|f\|\cdot\|g\|.$$

Hence, as we have proved earlier the multiplication $\hat{\times}$ is commutative and there is a linear and $\hat{\times}$-multiplicative functional $\tilde{F}$ on A such that $\|\tilde{F} - F\| \leq O(\varepsilon)$. Notice also that if ε is sufficiently small then e is an invertible element of the algebra $(A,\hat{\times})$ and

$$f \times g = e^{-1} \hat{\times} f \hat{\times} g \qquad \text{for all } f,g \text{ in } A,$$

so the multiplication $\times$ is also commutative.

Put $F_1 : A \to \mathbb{C}$: $F_1(f) = \tilde{F}(\mathbb{1} \times f)$. We have

$$\|F_1 - F\| \leq \|F_1 - \tilde{F}\| + \|F - \tilde{F}\| = O(\varepsilon)$$

and

$$F_1(f) \cdot F_1(g) = \tilde{F}(\mathbb{1} \times f) \cdot \tilde{F}(\mathbb{1} \times g) = \tilde{F}((\mathbb{1} \times f) \hat{\times} (\mathbb{1} \times g)) =$$
$$= \tilde{F}(\mathbb{1} \times f \times g) = F_1(f \times g),$$

so F_1 is a linear and $\times$-multiplicative functional on A close to F.

●

Step 2. (F) ⇒ (iv) for function algebras.

Proof. Assume (F) is satisfied and denote by φ any map from ChA into $\mathfrak{M}(A_\times)$ such that

$$\|F - \varphi(F)\| \leq \varepsilon \qquad \text{for all } F \text{ in } ChA.$$

For any f,g in A we have

$$\|f \times g - f \cdot g\| = \sup_{F \in \partial A} |F(f \times g) - F(f \cdot g)| \leq$$

$$\leq \frac{1}{1-\varepsilon} \sup_{F \in ChA} |\varphi(F)(f \times g) - \varphi(F)(f \cdot g)| \leq$$

$$\leq \frac{1}{1-\varepsilon} \sup_{F \in ChA} \{|\varphi(F)(f)\varphi(F)(g) - F(f)F(g)| + |F(fg) - \varphi(F)(fg)|\} \leq$$

$$\leq 4\varepsilon \|f\| \cdot \|g\|.$$

●

Step 3. (iv) ⇒ (iii) + (e) for function algebras.

Proof. Implication (iv) ⇒ (iii) is just Proposition 2.3. To show (iv) ⇒ (e) notice that if $\times$ is an algebraic ε-perturbation of A then

$$\|\mathbb{1} \times f - f\| \leq \varepsilon \|f\| \qquad \text{for all } f \text{ in } A,$$

hence, if $\varepsilon < 1$ then the operator $T: A \to A: f \mapsto \mathbb{1} \times f$ is an isomorphism with $\|T^{-1}\| \leq 1 + \varepsilon$ so there is an element e of A such that $\|e\| \leq$ $\leq 1 + \varepsilon$ and $\mathbb{1} \times e = \mathbb{1}$. By Proposition 2.3 the multiplication $\times$ is commutative and a simple computation shows that e is the unit of the algebra $(A, \times)$. From (iv) for $f = e$ and $g = \mathbb{1}$ we get

$$\|e - \mathbb{1}\| \leq \varepsilon(1 + \varepsilon).$$

●

Step 4. (iii) ⇒ (i) for function algebras.

Proof. For the proof we need the following lemma.

4.1. Lemma. Let T be a continuous linear map from a function algebra A onto a function algebra B with $\|T\| = 1$ and $\|T^{-1}\| \leq 1 + \varepsilon <$ < 1.5 . Then there is a dense subset Ω of the Shilov boundary of A

such that for each x in Ω there is a y in the Shilov boundary of B such that

$$|Tf(y)| \geq |f(x)| - 2\varepsilon\|f\| \qquad \text{for all } f \text{ in } A. \tag{13}$$

•

Proof of the lemma. In the sequel we call a net (g_α), of elements of the function algebra B, a peaking net at a point $y \in \mathrm{Ch}B$ if

$$\|g_\alpha\| = g_\alpha(y) = 1 \qquad \text{for all } \alpha$$

and

$$(g_\alpha) \text{ tends uniformly to zero outside any neighbourhood of } y.$$

We shall denote by Ω_y the subset of ∂A consisting of all points x_o admitting a net $(g_\alpha) \subset B$ peaking at y and a net $(x_\alpha) \subset \partial A$ converging to x_o and such that

$$|T^{-1}g_\alpha(x_\alpha)| \geq 1 - \varepsilon \qquad \text{for all } \alpha.$$

Since $\|T\| = 1$ and because the set ∂A is compact it follows that the set Ω_y is non-void.

Notice now that for any f in A and for a suitable net (η_α) of complex numbers of modulus one we have

$$\overline{\lim_\alpha} \|f + \eta_\alpha T^{-1}(g_\alpha)\| \geq 1 - \varepsilon + |f(x_o)|,$$

hence

$$\overline{\lim_\alpha} \|T(f) + \eta_\alpha g_\alpha\| \geq (1 - \varepsilon + |f(x_o)|)\cdot(1 + \varepsilon)^{-1}.$$

Thus, by the definition of (g_α) we obtain

$$|Tf(y)| \geq (1 - \varepsilon + |f(x_o)|)(1 + \varepsilon)^{-1} - 1 \geq |f(x_o)| - 2\varepsilon$$

for any function f in A of norm one.

So we have proved that (13) is fulfilled for any y in $\mathrm{Ch}B$ and any x_o in Ω_y. It remains to prove that $\Omega_o = \bigcup\{\Omega_y : y \in \mathrm{Ch}B\}$ is a dense subset of ∂A. Suppose that it is not. Then there exists an open subset V of ∂A such that $\overline{V} \cap \Omega_o = \emptyset$. Fix $0 < \delta < 1 - 2\varepsilon$ and an element f_1 from A such that

$$\|f_1\| = 1 \qquad \text{and} \qquad |f_1(x)| < \delta \qquad \text{for} \qquad x \in \partial A \setminus V.$$

Take $y_1 \in \text{ChB}$ such that

$$|Tf_1(y_1)| \geq (1+\varepsilon)^{-1} > 1-\varepsilon,$$

and let $(g_\alpha) \subset B$ be a net peaking at y_1. For a suitable net (ξ_α) of complex numbers of modulus one we have

$$\overline{\lim_\alpha} \|Tf_1 + \xi_\alpha g_\alpha\| > 1-\varepsilon+1.$$

Hence

$$\overline{\lim_\alpha} \|f_1 + \xi_\alpha T^{-1}(g_\alpha)\| > 2-\varepsilon,$$

so by the definition of f_1, there exists a net $(x_\alpha) \subset V$ such that

$$\overline{\lim_\alpha} |T^{-1}(g_\alpha)(x_\alpha)| > 1-\varepsilon. \tag{14}$$

Because $\overline{V}$ is a compact subset of ∂A we can assume that the net (x_α) converges to $x_o \in \overline{V}$, which leads, in view of (14), to the conclusion that $x_o \in \overline{V} \cap \Omega_o$. But this contradicts the assumption $\overline{V} \cap \Omega_o = \emptyset$ and therefore proves the lemma.

●

Let us now return to the proof of the implication (iii) ⇒ (i). To this end let A and B be function algebras and let T be a continuous linear map from A onto B such that $\|T\| \leq 1+\varepsilon$, $\|T^{-1}\| \leq 1+\varepsilon$. Without loss of generality we can assume that $\|T\| = 1$. From the lemma we have for any f,g in A

$$\|Tf\cdot Tg\| = \sup_{y\in\partial B} |Tf(y)Tg(y)| \geq \sup_{x\in\Omega} (|f(x)| - 2\varepsilon\|f\|)\cdot(|g(x)| - 2\varepsilon\|g\|) \geq$$

$$\geq \|fg\| - 4\varepsilon\|f\|\cdot\|g\| + 4\varepsilon^2\|f\|\cdot\|g\|,$$

so

$$\|Tf\cdot Tg\| - \|fg\| \geq -4\varepsilon\|f\|\cdot\|g\| \quad (\text{if } \varepsilon < \tfrac{1}{2}).$$

Put now $T_1 = T^{-1}/\|T^{-1}\|$ and $f_1 = T_1^{-1}(f)$, $g_1 = T_1^{-1}(g)$. In the same manner as above we derive from the lemma

$$\|f\cdot g\| - \|T_1^{-1}(f)\cdot T_1^{-1}(g)\| = \|T_1(f_1)\cdot T_1(g_1)\| - \|f_1\cdot g_1\| \geq$$

$$\geq -4\varepsilon\|f_1\|\cdot\|g_1\| \geq -4\varepsilon(1+\varepsilon)^2\|f\|\cdot\|g\|.$$

Hence

$$\|f\cdot g\| - \|Tf\cdot Tg\| \geq -4\varepsilon(1+\varepsilon)^2\|f\|\cdot\|g\| - (\|Tf\cdot Tg\| +$$

$$- \|T_1^{-1}(f)\cdot T_1^{-1}(g)\|) \geq -7\varepsilon \|f\|\cdot\|g\| .$$

We get

$$| \|f\times g\| - \|f\cdot g\| | \leq 8\varepsilon \|f\|\cdot\|g\| \quad \text{for any } f,g \text{ in } A$$

where $f\times g = T^{-1}(Tf\cdot Tg)$.

•

<u>Step 5.</u> <u>(i) ⇒ (ii) for Banach algebras.</u>
We need the following lemma.

<u>4.2. Lemma</u>. Let A be a Banach space and let $\cdot, \times$ be two multiplications on A with unit elements $\mathbb{1}$ and e respectively. Assume that $(A,\cdot)$ is a Banach algebra and

$$| \|f\times g\| - \|f\cdot g\| | \leq \varepsilon \|f\|\cdot\|g\| \quad \text{for all } f,g \text{ in } A \tag{15}$$

then

$$\| \mathbb{1}^{-1}\times f\times g - f\cdot g \| \leq \varepsilon' \|f\|\cdot\|g\| \quad \text{for all } f,g \text{ in } A \tag{16}$$

where $\mathbb{1}^{-1}$ is the inverse element of $\mathbb{1}$ in the algebra $(A,\times)$ and ε and ε' tend to zero simultaneously.

•

<u>Proof of the lemma</u>. Replace f in (15) by the element $\exp(\lambda f)$ where λ is a complex number and $\exp$ is the exponential function in the Banach algebra $(A,\cdot)$, and replace g by $\exp(-\lambda f)\cdot g$. We get

$$| \|g\| - \|\exp\lambda f\times (\exp(-\lambda f)\cdot g)\| | \leq \varepsilon \|g\|\cdot\|\exp(-\lambda f)\|\cdot\|\exp\lambda f\| \leq$$

$$\leq \varepsilon \|g\| \exp(2|\lambda|\cdot\|f\|).$$

This gives

$$\|g\| (1+\varepsilon\exp(2|\lambda|\cdot\|f\|)) \geq \|\exp(\lambda f)\times(\exp(-\lambda f)\cdot g)\| =$$

$$= \|(\mathbb{1}+\lambda f/1!+\lambda^2 f^2/2!+\ldots)\times(g-\lambda fg/1!+\lambda^2 f^2 g/2!-\ldots)\| =$$

$$= \|\mathbb{1}\times g - (f\times g - \mathbb{1}\times(fg))\lambda/1! + (\ldots)\lambda^2/2! + \ldots \| .$$

Now let F be any linear functional on the Banach space A of norm one. We have

$$|F(\mathbb{1}\times g) - \lambda/1!F(f\times g - \mathbb{1}\times(fg)) + \lambda^2/2!F(\ldots) + \ldots| \leq$$

$$\leq \|g\|\cdot(\varepsilon \exp(2|\lambda|\cdot\|f\|) + 1).$$

This shows that the modulus of the entire function

$$\varphi(\lambda) = F(\mathbb{1}\times g) - \lambda/1!F(f\times g - \mathbb{1}\times (fg)) + \lambda^2/2!F(\ldots) - \ldots$$

on the unit disc $D = \{\lambda \in \mathbb{C} : |\lambda| < 1\}$ is not greater than $\|g\|\cdot(\varepsilon \exp 2\,\|f\| + 1)$, hence the first derivative at the point zero of this function has the modulus not greater than this constant too. Because F is an arbitrary functional of norm one, it follows that

$$\|f\times g - \mathbb{1}\times (f\cdot g)\| \leq \|g\|\cdot(\varepsilon \exp 2\,\|f\| + 1) \quad \text{for any } f,g \in A. \tag{17}$$

Fix elements f and g in A both of norm one, and let $\alpha = -\frac{1}{2}\log \varepsilon > 0$. From (17) we get

$$\|f\times g - \mathbb{1}\times (f\cdot g)\| = \|(\alpha f)\times (g/\alpha) - \mathbb{1}\times ((\alpha f)\cdot (g/\alpha))\| \leq$$

$$\leq \alpha^{-1}(\varepsilon \exp 2\alpha + 1) = -4/\log\varepsilon$$

for $0 < \varepsilon < 1$. This shows the existence of a positive number $\varepsilon_1 = \varepsilon_1(\varepsilon) = -4/\log\varepsilon$ such that $\varepsilon_1(\varepsilon) \to 0$ as $\varepsilon \to 0$ and

$$\|f\times g - \mathbb{1}\times (f\cdot g)\| \leq \varepsilon_1 \|f\|\cdot\|g\| \quad \text{for any } f,g \text{ in } A. \tag{18}$$

Now let us estimate the norm of the unit e of the algebra $(A,\times)$. From (15) we have

$$|\,\|\mathbb{1}\| - \|e\|\,| = |\,\|\mathbb{1}\times e\| - \|\mathbb{1}\cdot e\|\,| \leq \varepsilon \|e\|,$$

and hence

$$\|e\| \leq (1-\varepsilon)^{-1}. \tag{19}$$

Setting $f = g = e$ in (18) we find

$$\|e - \mathbb{1}\times (e\cdot e)\| \leq \varepsilon_1 \|e\|^2 \leq \varepsilon_1 (1-\varepsilon)^{-2}.$$

Which implies that if ε is sufficiently small then $\mathbb{1}$ is an invertible element of $(A,\times)$. Further if

$$\|e - x\| \leq \varepsilon_1 \|e\|^2 < (1+\varepsilon)^{-1}$$

then

$$\|x^{-1}\| \leq \|e\| + \sum_{n=1}^{\infty} \|\underbrace{(e-x) \times \ldots \times (e-x)}_{n \text{ times}}\| = \|e\| +$$

$$+ \sum_{n=1}^{\infty} (1+\varepsilon)^{n-1}(\varepsilon_1 \|e\|^2)^n \leq (1-\varepsilon)^{-1}(1-\varepsilon_1(1+\varepsilon)\|e\|^2)^{-1}$$

so that

$$\|\mathbb{1}^{-1}\| = \|(\mathbb{1} \times (e \cdot e))^{-1} \times (e \cdot e)\| \leq$$

$$\leq (1+\varepsilon)(1-\varepsilon)^{-3}(1-\varepsilon_1(1+\varepsilon)(1-\varepsilon)^{-2})^{-1} = c(\varepsilon)$$

provided $\varepsilon_1(1+\varepsilon)(1-\varepsilon)^{-2} < 1$. Moreover, for each f and g in A we have

$$\|\mathbb{1}^{-1} \times f \times g - f \cdot g\| \leq (1+\varepsilon)\|\mathbb{1}^{-1}\| \cdot \|f \times g - \mathbb{1} \times (f \cdot g)\| \leq$$

$$\leq (1+\varepsilon)^2 c(\varepsilon) \cdot \|f\| \cdot \|g\| = \varepsilon' \|f\| \cdot \|g\| .$$

●

To end the proof of Step 5 assume (15), put $\chi = \mathbb{1} \times \mathbb{1}$ and denote by $\hat{\times}$ the multiplication on A defined by

$$f \mathbin{\hat{\times}} g = \mathbb{1}^{-1} \times f \times g \quad \text{for all } f,g \text{ in } A.$$

By the lemma we have

$$\|f \mathbin{\hat{\times}} g - f \cdot g\| \leq \varepsilon' \|f\| \cdot \|g\| \quad \text{for all } f,g \text{ in } A. \tag{20}$$

Hence for any f,g in A we find

$$\|f \times g - \chi \cdot f \cdot g\| = \|f \times g - (\mathbb{1} \times \mathbb{1}) \cdot (f \cdot g)\| =$$

$$= \|\mathbb{1}^{-1} \times (\mathbb{1} \times \mathbb{1}) \times (\mathbb{1}^{-1} \times f \times g) - (\mathbb{1} \times \mathbb{1})(f \cdot g)\| \leq$$

$$\leq \|(\mathbb{1} \times \mathbb{1}) \mathbin{\hat{\times}} (f \mathbin{\hat{\times}} g) - (\mathbb{1} \times \mathbb{1})(f \mathbin{\hat{\times}} g)\| + \|(\mathbb{1} \times \mathbb{1})(f \mathbin{\hat{\times}} g) - (\mathbb{1} \times \mathbb{1})(fg)\| \leq$$

$$\leq \varepsilon' \|\mathbb{1} \times \mathbb{1}\| \cdot \|f \mathbin{\hat{\times}} g\| + \|\mathbb{1} \times \mathbb{1}\| \varepsilon' \|f\| \cdot \|g\| .$$

By (15) we have

$$\|\chi\| \leq \|\mathbb{1} \times \mathbb{1}\| \leq 1 + \varepsilon$$

and by (20)

$$\| f \hat{\times} g \| \le (1+\varepsilon') \, \|f\| \cdot \|g\|$$

so we get

$$\| f \times g - \chi \cdot f \cdot g \| \le \varepsilon' (1+\varepsilon) ((1+\varepsilon') + 1) \, \|f\| \cdot \|g\| = \qquad (21)$$

$$= c'(\varepsilon) \cdot \|f\| \cdot \|g\| \, .$$

Hence it remains to show that χ is an invertible element of $(A,\cdot)$ and to estimate the norm of χ^{-1}. As in the proof of the lemma 4.2 we can show that $\|e\| \le (1-\varepsilon)^{-1}$ so from (21) for $f = e$ and $g = \mathbb{1}$ we get

$$\| \mathbb{1} - \chi \cdot e \| = \| \mathbb{1} \times e - (\mathbb{1} \times \mathbb{1}) \cdot \mathbb{1} \cdot e \| \le c'(\varepsilon) \, \| e \| \le c'(\varepsilon) (1-\varepsilon)^{-1} =$$

$$= c''(\varepsilon)$$

hence if ε is sufficiently small then $\chi \cdot e$ is an invertible element of $(A,\cdot)$ and

$$\| (\chi \cdot e)^{-1} \| \le (1 - c''(\varepsilon))^{-1}$$

so

$$\| \chi^{-1} \| = \| e (\chi \cdot e)^{-1} \| \le \| e \| \cdot \| (\chi \cdot e)^{-1} \| \le (1 - c''(\varepsilon))^{-1} (1-\varepsilon)^{-1} .$$

●

<u>Step 6</u>. <u>(ii) ⇒ (i) ⇒ (v) for Banach algebras.</u>

<u>Proof</u>. Evident.

●

<u>Step 7</u>. <u>(ii) + (e) ⇒ (iv) for Banach algebras.</u>

<u>Proof</u>. Let us assume

$$\| \chi \cdot f \cdot g - f \times g \| \le \varepsilon \| f \| \cdot \| g \| \quad \text{for all } f, g \text{ in } A \qquad (22)$$

where $\chi \in A^{-1}$, $\|\chi\| \le 1+\varepsilon$, $\|\chi^{-1}\| \le 1+\varepsilon$ and

$$\| \mathbb{1} - e \| < \varepsilon . \qquad (23)$$

For $f = \chi^{-1}$ and $g = e$ we get

$$\| e - \chi^{-1} \| \le \varepsilon \| \chi^{-1} \| \cdot \| e \| \le \varepsilon (1+\varepsilon)^2$$

hence

$$\| 1\!\!1 - \chi^{-1} \| \leq \varepsilon + \varepsilon(1+\varepsilon)^2 = \varepsilon'$$

and consequently

$$\| 1\!\!1 - \chi \| \leq \varepsilon'(1-\varepsilon')^{-1} = \varepsilon'' .$$

Finally from (22) we get

$$\| f \cdot g - f \times g \| \leq \| \chi \cdot f \cdot g - f \times g \| + \| \chi \cdot f \cdot g - f \cdot g \| \leq$$

$$= \varepsilon \| f \| \cdot \| g \| + \varepsilon'' \| f \| \cdot \| g \| = (\varepsilon + \varepsilon'') \| f \| \cdot \| g \|$$

for all f,g in A.

•

Step 8. (iv) ⇒ (F) for real function algebras.

Proof. Since any real function algebra can be identified with an algebra of the form $C_R(S)$ for some compact Hausdorff space S it is sufficient to prove the following proposition.

4.3.Proposition. Let S be a compact Hausdorff space and let $\times$ be an ε-perturbation of the algebra $C_R(S)$ then for all s_o in S there is a measure $\Delta\mu_{s_o}$ on S, of variation not greater than 5ε such that the functional on $C_R(S)$ which is represented by the measure $\mu = \delta_{s_o} + \Delta\mu_{s_o}$ is in the Shilov boundary of $(A,\times)$.

Proof. Fix $s_o \in S$, let $(U_\alpha)_{\alpha\in\Gamma}$ be a net of open neighbourhoods of s_o, and let $(f_\alpha)_{\alpha\in\Gamma}$ be a net of elements of $C_R(S)$ such that

$$\| f_\alpha \| = 1 = f(s_o), \quad f_\alpha(s) \geq 0 \quad \text{for } s \text{ in } S$$

and (24)

$$f_\alpha(s) = 0 \quad \text{for } s \in S \setminus U_\alpha \quad \text{for all } \alpha \in \Gamma.$$

By Proposition 2.2 for each $\alpha \in \Gamma$ there is a measure μ_α on S, which represents a functional from $\partial A_\times$, such that

$$|\mu_\alpha(f_\alpha)| \geq 1 - \varepsilon \quad \text{and} \quad \mathrm{var}(\mu_\alpha) \leq 1 + \varepsilon. \tag{25}$$

The inequality $\|1\!\!1 \times f - f\| \leq \varepsilon \|f\|$ provides that

$$|\mu_\alpha(1\!\!1) - 1| \cdot |\mu_\alpha(f)| = |\mu_\alpha(1\!\!1)\mu_\alpha(f) - \mu_\alpha(f)| =$$

$$= |\mu_\alpha(1\!\!1 \times f) - \mu_\alpha(f)| = |\mu_\alpha(1\!\!1 \times f - f)| \leq \varepsilon \|\mu_\alpha\| \cdot \|f\|$$

for all f in $C_R(S)$.

Hence $|\mu_\alpha(1\!\!1) - 1| \leq \varepsilon$ and from (24) and (25) we get

$$\mu_\alpha(U_\alpha) \geq 1 - 2\varepsilon \ . \tag{26}$$

Taking a net finer than $(\mu_\alpha)_{\alpha\in\Gamma}$, and using the weak $*$ compactness of $\partial A_\times$, we can assume without loss of generality, that $(\mu_\alpha)_{\alpha\in\Gamma}$ converges to $\mu_o \in \partial A_\times$. From (25) and (26) it follows that μ_o is of the form

$$\mu_o = \delta_{s_o} + \Delta\mu \quad \text{where} \quad \mathrm{var}(\Delta\mu) \leq 5\varepsilon .$$

●

<u>Step 9</u>. <u>$(F) \Rightarrow (\varphi)$ for function algebras</u>.

<u>Proof</u>. Let A be a function algebra, $\times$ any multiplication on A and assume that for any F in ChA there is an $\tilde{F}$ in $\mathfrak{M}(A_\times)$ such that $\|F - \tilde{F}\| \leq \varepsilon$. In other words assume that for any $s_o \in$ ChA there is a measure $\Delta\mu_o$ on ∂A of variation not greater than ε such that the measure $\delta_{s_o} + \Delta\mu_o$ represents a functional from $\mathfrak{M}(A_\times)$.

Denote by X the subset of $\mathfrak{M}(A_\times)$ consisting of all functionals which can be represented by a measure on ∂A of the form $\delta_{s_o} + \Delta\mu$ with $s_o \in$ ChA and $\mathrm{var}(\Delta\mu) < \varepsilon$. Let us define another norm $|||\cdot|||$ on A by

$$|||f||| = \sup\{|x(f)| : x \in X\}.$$

We have

$$(1 - \varepsilon)\|\cdot\| \leq |||\cdot||| \leq (1 + \varepsilon)\|\cdot\|$$

moreover $(A, \times, |||\cdot|||)$ is a function algebra which can be regarded as an algebra of continuous functions on some compactification $\bar{X}$ of the set X. (In fact, in the notation of Theorem $(A, \times, |||\cdot|||) = B$).

Notice that if $s_1, s_2 \in$ ChA, $s_1 \neq s_2$ then $\|s_1 - s_2\| = 2$ so if the measures $\delta_{s_1} + \Delta\mu_1$ and $\delta_{s_2} + \Delta\mu_2$ represent the same functional on A

and $\operatorname{var}(\Delta\mu_i) \leq \varepsilon < 1$, $i = 1,2$ then $s_1 = s_2$. So we can define a function φ from X onto $\mathrm{Ch}A$:

$$\varphi : X \to \mathrm{Ch}A : \varphi(x) = s \quad \text{if there is a measure } \Delta\mu \text{ on } \partial A$$

$$\text{with } \operatorname{var}(\Delta\mu) \leq \varepsilon \text{ and } x \equiv \delta_s + \Delta\mu .$$

The condition (φ) of Theorem is evidently fulfilled so it remains to prove that $X = \mathrm{Ch}A_x$ and φ is a homeomorphism.

a) φ is continuous.

To show this let $(x_i)_{i\in I}$ be a net of elements of X converging to $x_0 \in X$ and let $\varphi(x_i) = s_i \in \mathrm{Ch}A$,

$$x_i \equiv \delta_{s_i} + \Delta\mu_i, \ \operatorname{var}(\Delta\mu_i) \leq \varepsilon \quad \text{for} \quad i \in I \cup \{0\}.$$

Without loss of generality we can assume that $(s_i)_{i\in I}$ tends to $s_1 \in \partial A$ and $(\Delta\mu_i)_{i\in I}$ tends in the weak * topology to $\Delta\mu_1$. Hence the measures $\delta_{s_0} + \Delta\mu_0$ and $\delta_{s_1} + \Delta\mu_1$ represent the same functionals on A and since $s_0 \in \mathrm{Ch}A$ and $\operatorname{var}(\Delta\mu_1) \leq \underline{\lim}\operatorname{var}(\Delta\mu_i) \leq \varepsilon < 1$ so we get $s_0 = s_1$.

b) φ is a closed map.

To prove b) we need to show the following:

Let $(x_i)_{i\in I}$ be a net contained in X, let $\varphi(x_i) = s_i \in \mathrm{Ch}A$, $x_i \equiv \delta_{s_i} + \Delta\mu_i$ with $\operatorname{var}(\Delta\mu_i) \leq \varepsilon$ and let $s_i \to s_0 \in \mathrm{Ch}A$ then if $x_i \to x_0 \in \partial A$ then $x_0 \in X$ and $\varphi(x_0) = s_0$.

Without loss of generality we can assume that $(\Delta\mu_i)_{i\in I}$ tends in the weak * topology to $\Delta\mu_0$ and that $(\delta_{s_i} + \Delta\mu_i)_{i\in I}$ tends to $\delta_{s_0} + \Delta\mu_0$ and this means that the functional $x_0 \in \mathfrak{M}(A_x)$ can be represented by the measure $\delta_{s_0} + \Delta\mu_0$ with $\operatorname{var}(\Delta\mu_0) \leq \underline{\lim}\operatorname{var}(\Delta\mu_i) \leq \varepsilon$ so $x_0 \in X$ and $\varphi(x_0) = s_0$.

c) φ is one to one.

To show this fix $s_0 \in \mathrm{Ch}A$ and let $K = \varphi^{-1}(s_0)$. We need to demonstrate that K consists only of a single point. Notice that by the definition of φ we have

If $\varphi(x_1) = \varphi(x_2)$ then $|x_1(f) - x_2(f)| \leq 2\varepsilon \| f \|$ (*)

for all f in A,

so to end the proof of c) we show the following.

If K consists of more than one point then (**)

$$\inf\{ \| f \| : f \in \mathfrak{B} \} \leq 2$$

where

$\mathfrak{B} = \{ f \in A :$ there are $x_1, x_2 \in \bar{K}$ such that $x_1(f) = 0$

and $x_2(f) \geq 1 \}$.

To prove (**) it is sufficient to show that for all $f \in \mathfrak{B}$ there is an $h \in \mathfrak{B}$ such that $\|h\| \leq \max\{2, \frac{\|f\|}{2}\}$. For this purpose let $f \in \mathfrak{B}$, $x_1(f) = 0$, $x_2(f) = 1$ and $\sup\{|x(f)| : x \in K\} < 1 + \varepsilon$. Let U be a neighbourhood of K such that $|x(f)| < 1 + \varepsilon$ for $x \in U$ and let $g \in A$ be such that $g(s_o) = 1 = \| g \|$, $|g(s)| < \varepsilon$ for $s \in \partial A \setminus \varphi(X \setminus U)$ (such a g exists provided $s_o \in \mathrm{Ch}A$ and φ is a closed map). We have

$$x_1(f \times g) = x_1(f) \cdot x_1(g) = 0.$$

$$x_2(f \times g) - 1 = |x_2(g) - 1| \leq \varepsilon,$$

$$|f \times g(x)| \leq (1 + \varepsilon) |||g||| \leq (1 + \varepsilon)^2 \quad \text{for} \quad x \in U$$

and

$$|f \times g(x)| \leq |||f||| \sup_{x \notin U}\{ |g(\varphi(x)) + \Delta\mu_x(g)| \} \leq 2(1 + \varepsilon)\varepsilon \| f \| \quad \text{for} \quad x \in X \setminus U.$$

The desired result is now obtained by substituting $h = f \times g - (1 - x_2(f \times g))\mathbb{1}$ (for $\varepsilon < 0.2$).

It remains to prove that $X = \mathrm{Ch}A$ and for this we need the following the so-called "$\frac{3}{4} - \frac{1}{4}$" Bishop's criterion (see e.g. Larsen [1] page 236).

<u>Theorem (Bishop)</u>. Let A be a function algebra on a compact Hausdorff space X. A point $x_o \in X$ is in the Choquet boundary of A if and only if there are constants $K \geq 1$ and $c < 1$ such that for any open neighbourhood U of x_o there is an f in A such that

$$\| f \| \leq K, \quad f(x_o) = 1 \quad \text{and} \quad |f(x)| \leq c \quad \text{for} \quad s \in S \setminus U.$$

•

Fix now $x_o \in X$; by the assumption $s_o = \varphi(x_o)$ is a weak peak point for A so for any open neighbourhood U of s_o there is an $f \in A$ such that $f(s_o) = 1 = \|f\|$ and $|f(s)| \leq \varepsilon$ for $s \in \mathrm{Ch}A \setminus U$. On the other hand we have proved that φ is a homeomorphism and condition (φ) is satisfied, so for any open neighbourhood V of x_o there is an f in $(A, \times, |||\cdot|||)$ such that $|||f||| \leq 1+\varepsilon$, $|f(x_o) - 1| \leq \varepsilon$ and $|f(x)| \leq c\varepsilon$ for $x \in X \setminus V$ hence by the Bishop criterion x_o is a weak peak point for $(A, \times, |||\cdot|||)$ and this proves $X \subseteq \mathrm{Ch}A_{\times}$.

Consider now the function algebra $(A, \times, |||\cdot|||) \subset C(\bar{X})$. On this algebra we have also the same two multiplications $\times$ and $\cdot$, moreover

$$|||f \times g - f \cdot g||| \leq (1+\varepsilon)\, \|f \times g - f \cdot g\| \leq (1+\varepsilon)\varepsilon\, \|f\| \cdot \|g\| \leq$$

$$\leq \frac{(1+\varepsilon)\varepsilon}{(1-\varepsilon)^2} |||f||| \cdot |||g||| .$$

By the proved part of Theorem 3.1 condition (F) is fulfilled for the algebra $(A, \times, |||\cdot|||)$ and the second multiplication $\cdot$;this means that for any $G \in \mathrm{Ch}((A, \times, |||\cdot|||))$ there is a $\tilde{G} \in \mathfrak{M}(A)$ such that $\|G - \tilde{G}\| \leq \varepsilon'(\varepsilon)$. So by the same argument there is a subset Y of $\mathrm{Ch}A$ and a homeomorphism ψ from Y onto $\mathrm{Ch}((A, \times, |||\cdot|||))$ such that

$$|||Tf \circ \psi - f||| \leq \varepsilon'' |||f||| \quad \text{for all } f \text{ in } A.$$

Combining this and (φ) we get

$$\| f \circ \varphi \circ \psi - f\| \leq \varepsilon''' \| f \| \quad \text{for all } f \text{ in } A.$$

Hence, again by using the fact that $\|F_1 - F_2\| = 2$ if $F_1 \neq F_2 \in \mathrm{Ch}A$, we get $\varphi^{-1} = \psi$, and this proves $X = \mathrm{Ch}A$, and at the same time ends the proof of Theorem 3.1.

∎

<u>Remarks</u>. Small algebraic perturbations were first studied by R.V.Kadison, D.Kastler [1] and J.Phillips [1] and in a general situation by B.E.Johnson [2] , I.Raeburn and J.L.Taylor [1] and R.Rochberg [1-6]. B.E.Johnson and independently I.Raeburn and J.L.Taylor proved that if the second and the third cohomology groups of A vanish then any sufficiently small perturbation produces a new algebra which is algebraically isomorphic to the original one. R.Rochberg [5] using another methods proved part (iii) ⇒ (iv) ⇒ (φ) of Theorem 3.1 with the

additional assumptions $A = ChA$ and that each point of ∂A is G_δ. Furthermore R.Rochberg investigated algebras of analytic functions defined on finitely bordered Riemann surfaces from the point of view of connection between the small deformations of algebras and the small deformations of conformal structure of these Riemann surfaces (see Chapter V).

Theorem 3.1. part (i) $\Rightarrow$ (ii) $\Rightarrow$ (iii) $\Rightarrow$ (iv) is a modification of Theorems 1 and 4 from author's paper [2]. Theorem 3.1 part (iv) $\Rightarrow$ (v) was proved in author's paper [3] and the existence of the homeomorphism φ was established under the additional assumption "$\partial A = ChA$" in author's paper [1].

The name "a small metric perturbation" is new but the problem has a long history reaching back to the classical Banach-Stone theorem. In 1965 D.Amir [1] and independently M.Cambern [1] proved that the Banach-Stone theorem is stable i.e. if the Banach-Mazur distance between $C(S)$ and $C(S')$ is less than 2 then S and S' are homeomorphic. B.Cengiz [1] generalized this result to extremely regular subspaces of $C(S)$ and $C(S')$, respectively, and H.B.Cohen [1] showed that the constant 2 could not be improved.

II. INTO-ISOMORPHISMS

§ 5. Definitions and examples.

Let A, B be function algebras and let T be a continuous linear map from A into B which preserves units of the algebras. The main theorem of the first chapter states, roughly speaking, that if T is an onto map then T is an almost isometry if and only if T is an almost algebra isomorphism which takes place if and only if T is an almost composition with a homeomorphism of the Choquet boundaries of the algebras.

In this part we investigate into isomorphisms from the same point of view. Let us first consider two examples.

5.1. Example. Let S be any, non-one point, compact Hausdorff space and let P be any norm one, non-isometric map from C(S) into C(S) such that $P\mathbf{1} = \mathbf{1}$. Let $A = C(S)$, $B = C(S) \oplus C(S) = C(S \cup S)$ and define a map $T : A \to B : Tf = (f, Pf)$. The map T is an isometric embedding of A into B but evidently T need not be an algebra isomorphism though it becomes one if we compose it with a projection on the first component of B.

•

5.2. Example. Let A be the disc algebra, i.e. the algebra of all functions continuous on $D = \{z \in \mathbb{C} : |z| \leq 1\}$ which are holomorphic on intD, let S be a compact subset of D and let B be a completion of the algebra $A|_S$. We define T as follows

$$T : A \to B : Tf = f|_S .$$

The map T is a norm one algebra isomorphism, but if S does not contain $\partial A = \mathrm{Fr}\, D$ then T is not an isometric embedding, the image of T is not even closed in B.

•

The above examples show that, in general, there is no equivalence between the into isometries and the into algebra isomorphisms, but they also suggested that there is some connection. Notice also that since any function algebra can be isometrically embedded in an algebra C(S) so if we consider into isomorphisms, we can, without loss of generality, consider only maps from a function algebra A into C(S).

Since for into isomorphisms the definition of an ε-almost algebra isomorphisms from Chapter I could not be applied (since $T^{-1}(Tf\cdot Tg)$ need not be well-defined), in the sequel we need a more general definition.

5.3. Definition. A linear map T from a Banach algebra A into $C(S)$ is called an ε-isomorphism if

$$\|Tf\cdot Tg - T(f\cdot g)\| \leq \varepsilon\cdot\|f\|\cdot\|g\| \quad \text{for all } f,g \text{ in } A.$$

•

5.4. Definition. A linear map T from a Banach space A into B is called an ε-embedding if

$$(1+\varepsilon)\cdot\|f\| \geq \|Tf\| \geq (1-\varepsilon)\cdot\|f\| \quad \text{for all } f \text{ in } A.$$

•

By the definition any ε-embedding is a continuous map with norm not greater than $1+\varepsilon$. It turns out that the same is true for ε-isomorphism.

5.5. Proposition. Let A be a Banach algebra, let S be a compact Hausdorff space and let T be an ε-isomorphism from A into $C(S)$, then $\|T\| \leq 1+\varepsilon$.

Proof. To prove that $T: A \to C(S)$ is continuous with $\|T\| \leq 1+\varepsilon$ it is sufficient to show that for any s in S the functional T_s: $: A \to \mathbb{C} : T_s(f) = Tf(s)$ is continuous with $\|T_s\| \leq 1+\varepsilon$. The functional T_s is also an ε-isomorphism so without loss of generality we can assume that $C(S) = \mathbb{C}$. Notice next that if A does not possess a unit then we can extend T to $A \oplus \{\lambda e\}$ by putting $T(f+\lambda e) = T(f) + \lambda$ and the extended T is still an ε-isomorphism so without loss of generality we can also assume that A has a unit.

In the first step we show that T is a continuous map. If T is discontinuous then $\ker T$ is a dense subset of A so there is an f_0 in $\ker T$ with $\|f_0\| = 1$, $\|f_0^{-1}\| \leq 2$. Put $\|T(f_0^{-1})\| = c$. For any g in A we have

$$|T(g)| \leq |T(g\cdot f_0)T(f_0^{-1})| + |T(g - T(g\cdot f_0)f_0^{-1})| \leq$$

$$= c|T(g\cdot f_0) - T(g)\cdot T(f_0)| + |T(g\cdot f_0\cdot f_0^{-1}) - T(g\cdot f_0)T(f_0^{-1})| \leq$$

$$\leq c\varepsilon\|g\| + \varepsilon\|g\cdot f_0\|\cdot\|f_0^{-1}\| \leq (2+c)\varepsilon\|g\|,$$

so T is continuous.

To end the proof let g be any element of A with $\|g\| \leq 1$. We have

$$|T(g^2) - (T(g))^2| \leq \varepsilon$$

hence

$$\|T\| \geq |T(g^2)| \geq |T(g)|^2 - \varepsilon$$

and consequently

$$\|T\| \geq \|T\|^2 - \varepsilon$$

and this proves $\|T\| \leq 1 + \varepsilon$.

5.6. Remark. Let T be a linear isomorphism from a function algebra A onto a function algebra B such that

$$\|T^{-1}(Tf\cdot Tg) - f\cdot g\| \leq \varepsilon \|f\|\cdot\|g\| \qquad \text{for } f,g \in A.$$

Hence, by Proposition 5.5 we have $\|T\| \leq 1 + \varepsilon$ and so we get

$$\|Tf\cdot Tg - T(f\cdot g)\| \leq \|T\|\cdot\|T^{-1}(Tf\cdot Tg) - f\cdot g\| \leq$$

$$\leq \varepsilon(1+\varepsilon)\cdot\|f\|\cdot\|g\| \qquad \text{for } f,g \in A.$$

The above consideration proves that, for any function algebras A, B, the class of small onto isomorphisms defined by Definition 3.11 is not bigger (and in fact smaller) than this defined by Definition 5.3.

§ 6. ε-embeddings and ε-isomorphisms.

Our main result concerning ε-embeddings and ε-isomorphisms is the following.

6.1. THEOREM. Let S be a compact Hausdorff space, let A be a function algebra and let T be a linear map from A into $C(S)$ then:

a) if T is an ε-embedding with $\frac{1+\varepsilon}{1-\varepsilon} = k < 2$ then there is a subset S_1 of S and a continuous function φ from S_1 onto ChA such that

$$|Tf(s) - T\mathbb{1}(s) f\circ\varphi(s)| \le 4k\varepsilon\cdot\|T\|\cdot\|f\| \quad \text{for all } f \text{ in } A, \text{ and all } s \text{ in } S_1. \qquad (27)$$

Hence, if $T\mathbb{1} = \mathbb{1}$, then $\pi_{S_1}\circ T : A \to C(S_1)$ is a $16k\varepsilon$-isomorphism, moreover if $ChA = \partial A$ then φ can be extended to a continuous function from $\overline{S_1}$ onto ∂A.

b) if T is a norm increasing ε-isomorphism then T is continuous with $\|T\| \le 1+\varepsilon$, so it is an ε-embedding.

Proof of a).

Notice that taking $T/\|T\|$ in place of T we can assume without loss of generality that $\|T\| = 1$ and $\|T^{-1}\| = k < 2$.

Let us first introduce some notation. For each $x \in \partial A$ ($s \in S$) let δ_x (δ_s) denote, as usual, the evaluation map on A (on $C(S)$); let μ_x (ν_s) be any regular Borel measure on S (on ∂A) such that

$$\mu_x(Tf) = f(x) \qquad (\nu_s(f) = Tf(s)) \qquad \text{for all } f \in A$$

and

$$\mathrm{var}(\mu_x) = \|(T^{-1})^*\delta_x\| \qquad (\mathrm{var}(\nu_s) = \|T^*\delta_s\|).$$

Fix any positive number M such that $1/k > M > \max(\frac{1}{2}, \frac{1}{k} - \varepsilon^2)$ and for $x \in \partial A$ denote by $\tilde{S}_x$ the set of all points $s \in S$ such that $|\nu_s(\{x\})| \ge M\|T^*\delta_s\|$, and by S_x the set of all points s from $\tilde{S}_x$ such that $\|\delta_{s|TA}\|$ (i.e. the norm of the functional δ_s restricted to TA) is not less than M.

Notice that, because $M > 1/2$, we have $\tilde{S}_{x_1} \cap \tilde{S}_{x_2} = \emptyset$ if $x_1 \ne x_2$, so we can put $S_1 = \bigcup_{x\in ChA} S_x$ and define

$$\varphi : S_1 \to ChA : \varphi(s) = x \quad \text{if} \quad s \in S_x.$$

•

6.2. Lemma. For each x in ChA we have $S_x \ne \emptyset$ and:

(i) for any compact subset K of A the set $\bigcup_{x\in K} \tilde{S}_x$ is a closed subset of S.

(ii) $\varphi : S_1 \to ChA$ is continuous.

•

Proof. Let us first consider some general facts. If A is a closed subspace of $C(X)$ then any functional $F \in A^*$ can be extended, with the same norm, to a regular Borel measure μ_F on X. On the other hand A can be regarded as a subspace of the space of continuous functions on A^* with the weak $*$ topology, and any measure μ_F can be regarded as a regular Borel measure defined on a closed subset of A^*. Notice that this is independent of the fact as to whether or not A separates the points of set X. Now, let T be an isomorphism between some subspaces A and B of the spaces $C(X)$ and $C(S)$, respectively. Regarding A and B in the above way, for any functional $F \in A^*$ we can first find a measure μ_G on $S \subset B^*$ where $G = (T^{-1})^* F$ and then a measure ξ_G defined on $T^*(S) \subset A^*$:

$$\xi_G(K) = \mu_G((T^{-1})^*(K)) \quad \text{for any Borel subset } K \text{ of } A^*.$$

The measure ξ_G represents the same functional F on A and $\mathrm{var}(\xi_G) = \mathrm{var}(\mu_G) = \| (T^{-1})^*(F) \|$.

In the sequel we shall use the same sign for a function $f \in A \subset C(X)$ and for the corresponding continuous function on A^* if no misunderstanding is likely to occur.

We return now to our situation and put $F = \delta_{x_o}$. We get a measure $\xi_{x_o} = \xi_{(T^{-1})^* \delta_{x_o}}$ concetrated in $T^*(S) \subset A^*$ and such that

$$f(x_o) = \delta_{x_o}(f) = (T^{-1})^*(\delta_{x_o})(Tf) = \mu_{x_o}(Tf) = \xi_{x_o}(f) \tag{28}$$

for all f in A, and

$$\mathrm{var}(\xi_{x_o}) = \| (T^{-1})^* \delta_{x_o} \| = \mathrm{var}(\mu_{x_o}). \tag{29}$$

Fix $\varepsilon_o > 0$. By the definition of μ_{x_o} we have

$$\mu_{x_o}(\{s \in S : \| \delta_{s|TA} \| < 1\}) = 0$$

so by the regularity of μ_{x_o}, there is a compact subset S' of $\{s \in S : \|\delta_{s|TA}\| > M\}$ such that $|\mu_{x_o}(S \setminus S')| < \varepsilon_o$. From (28) and (29), for any f in A such that $\|f\| = 1 = f(x_o)$ there exists an $F \in \operatorname{supp} \xi_{x_o} \cap T^*(S')$ with the property that

$|F(f)| \geq 1 - 2\varepsilon_o$.

Let now κ be a continuous function on $D = \{z \in \mathbb{C} : |z| \leq 1\}$ which is holomorphic on $\mathrm{int}D$ and such that

$$\kappa : D \to D, \quad \kappa(0) = 0, \quad \kappa(1) = 1,$$

$|(\mathrm{Re}\,\kappa(z))^+ - \kappa(z)| \leq \varepsilon_o$ for all z in D, where for a real number t we define $t^+ = \max\{t, 0\}$. Let $\delta > 0$ be such that

$$|\kappa(z)| < \varepsilon_o \text{ for any } z \in \mathbb{C} \text{ with } |z| < \delta.$$

Let $(U_\alpha)_{\alpha \in \Gamma}$ be the net of open neighbourhoods of $x_o \in \mathrm{Ch}A$ and let $(\tilde{f}_\alpha)$ be a net of functions from A, such that for all $\alpha \in \Gamma$

$$\|\tilde{f}_\alpha(x_o)\| = 1 = \tilde{f}_\alpha(x_o)$$

and

$$|\tilde{f}_\alpha(x)| < \delta \text{ for all } x \text{ in } X \setminus U_\alpha.$$

Put $f_\alpha = \kappa \circ \tilde{f}_\alpha$. We have $f_\alpha \in A$ and

1. $\|f_\alpha(x_o)\| = 1 = f_\alpha(x_o)$;
2. $|f_\alpha(x)| < \varepsilon_o$ for all $x \in \partial A \setminus U_\alpha$;
3. $|f_\alpha(x) - (\mathrm{Re} f_\alpha(x))^+| < \varepsilon_o$ for all $x \in \partial A$.

Let $F_\alpha = T^*(\delta_{s_\alpha})$ be any functional such that $s_\alpha \in S'$ and

$$|F_\alpha(f_\alpha)| \geq 1 - 2\varepsilon_o. \tag{30}$$

Let us decompose the measure ν_{s_α} as follows:

$$\nu_{s_\alpha} = \nu_\alpha^1 + \nu_\alpha^2 \quad \text{where} \quad \nu_\alpha^1 = \nu_{s_\alpha}|_{U_\alpha}, \quad \nu_\alpha^2 = \nu_{s_\alpha}|_{X \setminus U_\alpha}.$$

We can assume that the nets (ν_α^1) and (ν_α^2) are weak $*$ convergent to measures ν^1 and ν^2, respectively, and (s_α) converges to $s_o \in S'$.

From (30) and the definition of ν_{s_α} and f_α we have

$$\mathrm{Re}(\nu_{s_\alpha}(U_\alpha)) \geq 1 - 6\varepsilon_o. \tag{31}$$

Hence $\nu^1 = \lambda_1 \delta_{x_o}$ where $|\lambda_1| \geq 1 - 6\varepsilon_o$. On the other hand the measure ν^2 can be represented in the form

$$\nu^2 = \lambda_2 \delta_{x_o} + \Delta\nu^2 \quad \text{where} \quad \Delta\nu^2(\{x_o\}) = 0 .$$

Hence

$$\nu^0 = \nu^1 + \nu^2 = (\lambda_1 + \lambda_2)\delta_{x_o} + \Delta\nu^2 .$$

From (31)

$$\mathrm{Re}(\nu_{s_\alpha}(U_\alpha)) \geq 1 - 6\varepsilon_o \geq \frac{1-6\varepsilon_o}{k} (|\nu_{s_\alpha}|(U_\alpha) + \mathrm{var}(\nu_\alpha^2)) .$$

Hence, if $1-6\varepsilon_o \geq M \cdot k$ then

$$|\lambda_1| \geq M(|\lambda_1| + \mathrm{var}(\nu^2)) = M(|\lambda_1| + |\lambda_2| + \mathrm{var}(\Delta\nu^2)) . \tag{32}$$

Notice now that, since $\frac{1}{2} < M < 1$ then for any $\lambda_1, \lambda_2 \in \mathbb{C}$ we have

$$|\lambda_1 + \lambda_2| - |\lambda_1| \geq M(|\lambda_1 + \lambda_2| - |\lambda_1| - |\lambda_2|) . \tag{33}$$

Adding by sides (32) and (33) we get

$$|\lambda_1 + \lambda_2| \geq M(|\lambda_1 + \lambda_2| + \mathrm{var}(\Delta\nu^2)) \geq M \ \mathrm{var} \ (\nu^0) \geq M \, \|T^*\delta_{s_o}\| .$$

Let now ν be any regular measure on ∂A which represents $T^*\delta_{s_o}$. The point x_o being in the Choquet boundary of A we have $|\nu(\{x_o\})| \geq$ $\geq |\lambda_1 + \lambda_2|$, hence by the foregoing inequality

$$|\nu(\{x_o\})| \geq M \| T^*\delta_{s_o} \|$$

and this proves that $s_o \in S_{x_o}$.

(i) Fix now a compact subset K of $\mathrm{Ch}A$ and let $s_\alpha \in \tilde{S}_{x_\alpha}$, $x_\alpha \in K$ and $s_\alpha \to s_o \in S$.

Recall that ν_{s_α} denotes a regular, Borel measure on S which represents the functional $T^*\delta_{s_\alpha}$ and is such that $\mathrm{var}(\nu_{s_\alpha}) = \|T^*\delta_{s_\alpha}\|$.

By our assumption

$$\nu_{s_\alpha} = \lambda_\alpha \delta_{x_\alpha} + \Delta_{\nu_\alpha} \quad \text{where} \tag{34}$$

$$|\lambda_\alpha| \geq M(|\lambda_\alpha| + \operatorname{var}(\Delta\nu_\alpha)).$$

Without loss of generality we can assume that the nets $(\lambda_\alpha),(x_\alpha)$ and $(\Delta\nu_\alpha)$ are convergent, in appropriate topologies, to $\lambda_o \in \mathbb{C}$, $x_o \in K$ and to a measure $\Delta\nu_o = \lambda'\delta_{x_o} + \rho$, where $\rho(\{x_o\}) = 0$, respectively.

From (34) we have

$$|\lambda_o| \geq M(|\lambda_o| + |\lambda'| + \operatorname{var}(\rho)),$$

hence, as in the proof of the previous part we get

$$|\lambda_o + \lambda'| \geq M(|\lambda_o + \lambda'| + \operatorname{var}(\rho)).$$

Next we get $|\nu_{s_o}(\{x_o\})| \geq M \operatorname{var}(\nu_{s_o})$ and hence $s_o \in \bigcup_{x \in K} \tilde{S}_x$.

(ii) Let $s_\alpha \in S_{x_\alpha}$ for $\alpha \in \Gamma$ and $s_\alpha \to s_o \in S_{x_o}$. To prove the continuity of φ it is sufficient to show that x_α is not convergent to any x_1 from the Shilov boundary of A with $x_1 \neq x_o$.

Assume that $x_\alpha \to x_1 \neq x_o$, $x_1 \in \partial A$, $x_o \in \operatorname{Ch}A$. We have

$$\nu_{s_\alpha} = \lambda_\alpha \delta_x + \Delta\nu_\alpha \quad \text{for } \alpha \in \Gamma \cup \{0\}$$

where $|\lambda_\alpha| \geq M(|\lambda_\alpha| + \operatorname{var}(\Delta\nu_\alpha))$.

Assume, without loss of generality, that $\lambda_\alpha \to \lambda'$, $\Delta\nu_\alpha \to \Delta\nu'$. $T^*\delta_{s_\alpha}$ tending to $T^*\delta_{s_o}$ in the weak $*$ topology it follows that the measures

$$\lambda'\delta_{x_1} + \Delta\nu' \quad \text{and} \quad \lambda_o\delta_{x_o} + \Delta\nu_o$$

represent the same functionals on A, but this is impossible since $|\lambda'| + |\lambda_o| > \operatorname{var}(\Delta\nu') + \operatorname{var}(\Delta\nu_o)$ and $x_o \in \operatorname{Ch}A$.

To end the proof of a) we have to verify (27), and to show, that if $\operatorname{Ch}A = \partial A$ then φ can be extended to a continuous function from $\overline{S_1}$ onto ∂A. To show this last assertion we put

$$\tilde{\varphi} : \bigcup_{x \in X} \tilde{S}_x \to \partial A : \tilde{\varphi}(s) = x \quad \text{if} \quad s \in \tilde{S}_x.$$

For a compact space $\operatorname{Ch}A = A$ the inclusion $\overline{S_1} \subset \bigcup_{x \in X} \tilde{S}_x$ and the continuity of $\tilde{\varphi}$ are immediate consequences of Lemma 6.2 (i).

To prove (27) fix f in A and s in $S_x \subset S_1$ with $\nu_s = \lambda\delta_x + \Delta\nu$ and $|\lambda| \geq M(|\lambda| + \mathrm{var}(\Delta\nu)) = M\,\mathrm{var}(\nu_s)$. We have

$$|Tf(s) - T\mathbb{1}(s) f\circ\varphi(s)| = |\lambda f(x) + \Delta\nu(f) - (\lambda + \Delta\nu(\mathbb{1}))f(x)| \leq$$

$$\leq 2\,\mathrm{var}(\Delta\nu)\cdot\|f\| \leq 2\,\|T^*\delta_s\|\cdot(1-M)\cdot\|f\| \leq 4k\varepsilon\cdot\|f\| .$$

•

Proof of b). An immediate consequence of Proposition 5.5.

•

Remarks. The source of the problems of this chapter is the classical Banach-Stone theorem. D.Amir [1] and M.Cambern [1] independently generalized this theorem by proving that if $C_o(S)$ and $C_o(S')$ are isomorphic under an isomorphism T satisfying $\|T\|\cdot\|T^{-1}\| < 2$ then S and S′ are homeomorphic. The first generalization of the Banach--Stone theorem to the theory of into isomorphisms is due to W.Holsztyński [1]; he proved that if there is a linear isometry of C(S) into C(S′) then S is a continuous image of a closed subset of S′. Y.Benyamini [1] found, for compact metric spaces, a common generalization of the theorems of Amir-Cambern and of Holsztyński, i.e. he proved Theorem 6.1 a) for metrizable set S. In author's paper [4] it is shown that the assumption "S is metric" can be omitted.

III. ISOMETRIES IN SEMISIMPLE, COMMUTATIVE BANACH ALGEBRAS

§ 7. Introduction.

In the first two chapters we considered the uniform algebras, and the one of our basic results was a generalization of the Nagasawa theorem stating that the metric and the algebraic perturbations of a function algebra produce the same class of algebras. One can not however expect that such a result will be true in general case since then even the Nagasawa theorem does not hold as the following easy example shows:

7.1. Example. Let $A = \{(a_j)_{j=0}^{\infty} : \sum_{j=0}^{\infty} |a_j| = \|(a_j)\| < \infty\}$, and define a multiplication on A by

$$(a_j)\cdot(b_j) = (c_j) \quad \text{where} \quad c_j = \sum_{i=0}^{j} a_i b_{j-i}.$$

We get a commutative, semisimple Banach algebra A with unit such that every isometry from A onto itself which preserves the unit is of the form

$$T((a_j)_{j=0}^{\infty}) = (a_0, \lambda_1 a_{\pi(1)}, \lambda_2 a_{\pi(2)}, \ldots),$$

where $|\lambda_j| = 1$ for $j = 1,2,\ldots$ and π is a permutation of positive integers. Any such isometry is an algebra isomorphism only if $\pi = id_{\mathbb{N}}$ and $\lambda_j = z^j$ for some $z \in \mathbb{C}$ with $|z| = 1$ and all $j = 1,2,\ldots$, so does not in general.

•

In this chapter we show however that there is a fairly large class of Banach algebras for which the Nagasawa theorem holds. The fact that the Nagasawa theorem does or does not hold for some Banach algebras A and B depends not only on the algebraic structures of A and B but also, and in fact primarily, on the norms in these algebras, and any Banach algebra A admits a number of equivalent and submultiplicative norms. We shall define natural norms on semisimple, commutative Banach algebras with unit; we show that any such algebra possesses a number of natural norms and we prove that the Nagasawa theorem holds for any semisimple, commutative Banach algebra with unit if it is equipped with a natural norm. For a large class of Banach algebras the natural norm coincides with a "usual" norm. The results of this chapter also hold for some normed spaces not being algebras. Before we formulate the result we need some definitions and notation.

§ 8. Definitions and notation.

We denote by $\mathbb{P}$ the set of all norms p on the two dimensional real linear space with $p((1,0)) = 1$.

For $p \in \mathbb{P}$ we put

$$D(p) = \lim_{t \to 0^+} \frac{p(1,t)-1}{t} .$$

For $z_o \in \mathbb{C}$ and $r \geq 0$ we put

$$K(z_o,r) = \{z \in \mathbb{C} : |z - z_o| \leq r\}$$

and we write $K(r)$ in place of $K(0,r)$.

If K,H are subsets of the complex plane $\mathbb{C}$ we denote by $co(K)$ the convex hull of K; we put

$$K + H = \{w + z : w \in K, z \in H\},$$

and for $z_o \in K$ we put

$$\rho(K,z_o) = \sup\{r \geq 0 : \exists z \in K \quad z_o \in K(z,r) \subset K\},$$

$$\rho(K) = \inf\{\rho(K,z) : z \in K\}.$$

Assume X is a compact, Hausdorff space, and A is a linear subspace of $C(X)$ and assume A contains the function constantly equal to one which we denote by $\mathbb{1}$, then:

- by $\|\cdot\|_\infty$ we denote the usual sup-norm on A;
- a seminorm $\|\cdot\|$ on A we call one-invariant if $\|a + \mathbb{1}\| = \|a\|$ for all a in A;
- let $p \in \mathbb{P}$ and let $\|\cdot\|$ be a norm on A, then we call it p-norm if there is a one-invariant seminorm $|||\cdot|||$ on A such that $\|\cdot\| = p(\|\cdot\|_\infty, |||\cdot|||)$;
- we call $\|\cdot\|$ a natural norm on A if it is a p-norm for some $p \in \mathbb{P}$;
- by $\bar{A}$ we denote the closure of A in $(C(X), \|\cdot\|_\infty)$;
- by ChA we denote the set of extreme points F of the unit ball of $(A, \|\cdot\|_\infty)^*$ such that $F(\mathbb{1}) = 1$ and we identify ChA with a

subset of X;

- for $f \in A$ we put

  $\sigma(f) = f(X)$ and $\tilde{\sigma}(f) = \mathrm{co}(f(X))$;

- we call A a regular subspace of $C(X)$ if for any $\varepsilon > 0$, any $x_o \in \mathrm{Ch}A$ and any open neighbourhood U of x_o there is an $f \in A$ with $\|f\|_\infty \leq 1+\varepsilon$, $f(x_o) = 1$ and $|f(x)| < \varepsilon$ for $x \in X \setminus U$.

In the sequel any semisimple, commutative Banach algebra A is identified, via the Gelfand transformation, with a subalgebra of $C(\mathfrak{M}(A))$ with $\mathfrak{M}(A)$ being the maximal ideal space of A.

§ 9. Isometries between natural algebras.

9.1. THEOREM. Let X, Y be compact Hausdorff spaces, let A and B be complex linear subspaces of $C(X)$ and $C(Y)$, respectively, and let $p, q \in \mathbb{P}$. Assume A and B contain constant functions and let $\|\cdot\|_A$, $\|\cdot\|_B$ be a p-norm and a q-norm on A and B, respectively. Assume next that there is a linear isometry T from $(A, \|\cdot\|_A)$ onto $(B, \|\cdot\|_B)$ with $T\mathbb{1} = \mathbb{1}$. Then if $D(p) = D(q) = 0$ or if A and B are regular subspaces of $C(X)$ and $C(Y)$, respectively, then T is an isometry from $(A, \|\cdot\|_\infty)$ onto $(B, \|\cdot\|_\infty)$.

Before proving the theorem let us prove two propositions which show that any semisimple, commutative Banach algebra A with unit possesses natural norms and is a regular subspace of $C(\mathfrak{M}(A))$,

9.2. Proposition. Let $(A, \|\cdot\|, \mathbb{1})$ be a commutative, semisimple Banach algebra with unit then for any $p \in \mathbb{P}$ there is a p-norm $\|\cdot\|_A$ on A which is submultiplicative and equivalent to the original one.

Proof. Fix $x_o \in \mathfrak{M}(A)$, put $A_o = \{a \in A : a(x_o) = 0\}$ and let P be a linear, continuous projection from A onto A_o with $P\mathbb{1} = 0$. Fix a positive integer k and define

$$\|a\|_A = p(\|a\|_\infty, k\|Pa\|) \quad \text{for } a \in A.$$

It is easy to check that $\|\cdot\|_A$ is a p-norm on A, equivalent to the original one and that if k is sufficiently large then $\|\cdot\|_A$ is submultiplicative.

•

9.3. Proposition. Let $(A, \|\cdot\|, \mathbf{1})$ be a commutative, semisimple Banach algebra with unit, then A is a regular subspace of $C(\mathfrak{M}(A))$.

Proof. Fix an $x_o \in \mathrm{Ch}A$. $(\bar{A}, \|\cdot\|_\infty)$ is a function algebra with the Choquet boundary equal to $\mathrm{Ch}A$, so there is a net $(f_\alpha)_{\alpha\in\Gamma} \subset \bar{A}$ such that $\|f_\alpha\| = 1 = f(x_o)$ and (f_α) tends uniformly to zero off any neighbourhood of x_o. Since $(A, \|\cdot\|_\infty)$ is a dense subset of $(\bar{A}, \|\cdot\|_\infty)$, for any $\varepsilon > 0$ there is an $\tilde{f}_\alpha \in A$ with $\|\tilde{f}_\alpha - f_\alpha\|_\infty < \varepsilon$ and this ends the proof.

•

Proof of Theorem 9.1. For any convex subset K of the complex plane and any $\varphi \in [0,2\pi)$ we put

$$c(K,\varphi) = \sup\{a \in \mathbb{R}: \text{there is a } b \in \mathbb{R} \text{ with } ae^{i\varphi} + be^{i\varphi + i\frac{\pi}{2}} \in K\}.$$

For any subspace A of $C(X)$ we define the following two functions

$$c_A : A \times [0,2\pi) \to \mathbb{R} \quad \text{by} \quad c_A(f,\varphi) = c(\tilde{\sigma}(f),\varphi),$$

$$r_A : A \times \mathbb{R}^+ \times [0,2\pi) \to \mathbb{R}^+ \quad \text{by} \quad r_A(f,t,\varphi) = \|f + e^{i\varphi}t\mathbf{1}\|_\infty.$$

For any $\varphi \in [0,2\pi)$, $f \in A$, $t \in \mathbb{R}^+$ we have

$$t + c_A(f,\varphi) \leq r_A(f,t,\varphi) \leq \sqrt{(t + c_A(f,\varphi)^2 + \|f\|_\infty^2}$$

and hence

$$\lim_{t\to+\infty} (r_A(f,t,\varphi) - t) = c_A(f,\varphi). \tag{35}$$

Let

$$\|\cdot\|_A = p(\|\cdot\|_\infty, \ |||\cdot|||_A);$$

$$\|\cdot\|_B = q(\|\cdot\|_\infty, \ |||\cdot|||_B),$$

where $|||\cdot|||_A$ and $|||\cdot|||_B$ are one-invariant seminorms on A and B respectively.

Fix $f \in A$ and put $c_1 = |||f|||_A$, $c_2 = |||Tf|||_B$. By our assumption for any positive t and any $\varphi \in [0,2\pi)$ we have

$$p(r_A(f,t,\varphi),c_1) = q(r_B(Tf,t,\varphi),c_2).$$

Hence, from (35) we get

$$0 = \lim_{t\to+\infty} [p(r_A(f,t,\varphi),c_1) - q(r_B(Tf,t,\varphi),c_2)] =$$

$$= \lim_{t\to+\infty} [r_A(f,t,\varphi)p(1, \frac{c_1}{r_A(f,t,\varphi)}) - r_A(f,t,\varphi)] +$$

$$+ \lim_{t\to+\infty} [r_B(Tf,t,\varphi) - r_B(Tf,t,\varphi)q(1,\frac{c_2}{r_B(Tf,t,\varphi)})] +$$

$$+ \lim_{t\to+\infty} (r_A(f,t,\varphi) - r_B(Tf,t,\varphi)) =$$

$$= c_1D(p) - c_2D(q) + c_A(f,\varphi) - c_B(Tf,\varphi).$$

Hence

$$c_B(Tf,\varphi) - c_A(f,\varphi) = |||\, f \,|||_A D(p) - |||\, Tf \,|||_B D(q) \tag{36}$$

for all $f \in A$ and $\varphi \in [0,2\pi)$.

<u>9.4. Lemma</u>. Let K and H be compact, convex subsets of the complex plane. Assume

$$c(K,\varphi) = c(H,\varphi) \quad \text{for all} \quad \varphi \in [0,2\pi)$$

then $K = H$.

•

<u>Proof of the lemma</u>. Assuming the contrary we get $z_o \in H \setminus K$. Since K is convex and compact there is a line l such that $\mathbb{C} \setminus l$ is a sum of two connected components $\mathbb{C}^+$ and $\mathbb{C}^-$ with $K \subset \mathbb{C}^+$ and $z_o \in \mathbb{C}^-$. Let $\varphi_o \in [0,2\pi)$ be such that the vector $(\cos\varphi_o, \sin\varphi_o)$ is orthogonal to l and has the direction from $\mathbb{C}^+$ to $\mathbb{C}^-$. We have

$$c(K,\varphi_o) < c(\mathbb{C}^+,\varphi_o) < c(\{z_o\},\varphi_o) \le c(H,\varphi_o).$$

For any $f \in A$ we denote $\Delta f = |||\, f \,|||_A D(p) - |||Tf|||_B D(q)$. For any $r \ge 0$ and for any compact convex subset K of $\mathbb{C}$ we have

$$c(K + K(r),\varphi) = c(K,\varphi) + r \quad \text{for all} \quad \varphi \in [0,2\pi).$$

Hence by (36) and the lemma for any $f \in A$ we get:

$$\left.\begin{array}{l} \text{if} \quad \Delta f \geq 0 \quad \text{then} \quad \tilde{\sigma}(Tf) = \tilde{\sigma}(f) + K(\Delta f); \\ \text{if} \quad \Delta f \leq 0 \quad \text{then} \quad \tilde{\sigma}(f) = \tilde{\sigma}(Tf) + K(-\Delta f), \end{array}\right\} \tag{37}$$

and hence

$$\| Tf \|_\infty - \| f \|_\infty = \Delta f. \tag{38}$$

Assume $D(p) = D(q) = 0$, then for any $f \in A$ we have $\Delta f = 0$ so (38) gives the thesis.

Assume now A and B are regular subspaces of $C(X)$ and $C(Y)$ respectively. To end the proof, by symmetry, it is sufficient to show that

$$\| Tf \|_\infty - \| f \|_\infty = \Delta f \geq 0 \quad \text{for all} \quad f \in A. \tag{39}$$

For this for any $\varepsilon > 0$ we put

$$\mathbf{A}_\varepsilon = \{ f \in A : \rho(\tilde{\sigma}(f)) \leq \varepsilon \} .$$

We shall now prove (39) by showing the following three statements:

1. T is continuous map from $(A, \|\cdot\|_\infty)$ onto $(B, \|\cdot\|_\infty)$.
2. For each $\varepsilon > 0$ the set $\mathbf{A}_\varepsilon$ is dense in $(A, \|\cdot\|_\infty)$.
3. For each $\varepsilon > 0$ and each $f \in \mathbf{A}_\varepsilon$ we have

$$\| Tf \|_\infty \geq \| f \|_\infty - \varepsilon .$$

<u>Proof of the first step</u>.

<u>9.5. Lemma</u>. Assume A is a regular subspace of $C(X)$ with $\mathbf{1} \in A$ and let $x_o \in \mathrm{Ch}A$ then for any $\varepsilon > 0$ and any open neighbourhood U of x_o there is an $f \in A$ such that

$$\| f \|_\infty \leq 1 + \varepsilon, \quad f(x_o) = 1, \quad |f(x) + 1| \leq \varepsilon \quad \text{for} \quad x \in X \setminus U$$

and $|\mathrm{Im} f(x)| \leq \varepsilon$ for all $x \in X$.

•

<u>Proof of the lemma</u>. We shall define, by induction, a sequence $(f_k)_{k=1}^\infty \subset A$ and a descending sequence of neighbourhoods $(U_k)_{k=1}^\infty$ of the point x_o:

We set $U_1 = U$ and let f_1 be any function from A such that $\| f_1 \|_\infty \leq 1 + \frac{\varepsilon}{2}$, $f_1(x_0) = 1$ and $|f_1(x)| < \frac{\varepsilon}{2}$ for all x in $X \setminus U_1$.

Assume that we have defined U_k and f_k then

$$U_{k+1} = \{x \in U_k : |f_k(x) - 1| < \tfrac{\varepsilon}{2}\},$$

$$\| f_{k+1} \|_\infty \leq 1 + \tfrac{\varepsilon}{2}, \quad f_{k+1}(x_0) = 1 \quad \text{and}$$

$$|f_{k+1}(x)| < \tfrac{\varepsilon}{2} \quad \text{for all } x \text{ in } X \setminus U_{k+1}.$$

Let n be a positive integer such that $\frac{1}{n} < \frac{\varepsilon}{6}$. We define

$$f = \frac{2}{n} \sum_{k=1}^{n} f_k - \mathbb{1}.$$

We have $f(x_0) = 1$, $|f(x) + 1| \leq \varepsilon$ for $x \in X \setminus U$. To end fix $x \in X \setminus U_1$ and denote by k_0 the greatest positive integer not greater than n such that $x \in U_{k_0}$. We find

$$|f(x)+1-2\cdot\frac{k_0-1}{n}| = |\frac{2}{n}[\sum_{j=1}^{k_0-1}(f_j(x)-1)+f_{k_0}(x)+\sum_{j=k_0+1}^{n} f_j(x)]|$$

$$\leq \frac{2}{n}[(k_0-1)\cdot\frac{\varepsilon}{2} + 1 + \varepsilon + (n-k_0)\frac{\varepsilon}{2}] = \frac{2}{n} + \frac{n+1}{n}\cdot\frac{\varepsilon}{2} < \varepsilon.$$

Hence we get $|\mathrm{Im} f(x)| \leq \varepsilon$ and next $\| f \|_\infty \leq 1+\varepsilon$.

•

To end the proof of this step fix $\varepsilon > 0$ and assume T is discontinuous. There is an $f_0 \in A$ such that $\| f_0 \|_\infty \leq \varepsilon$ and $\| Tf_0 \|_\infty = 1$. Since any function from B achieves its sup-norm on ChB we can assume there is a $y_0 \in ChB$ with $Tf_0(y_0) = 1$. We put $U_0 = \{y \in Y : |Tf_0(y)-1| < \varepsilon\}$. By the lemma there is a $g \in B$ such that $\| g \| \leq 1 + \varepsilon$, $g(y_0)=1$, $|g(y)+1| < \varepsilon$ for $y \in Y \setminus U$ and $|\mathrm{Im} g(y)| < \varepsilon$ for $y \in Y$. From (37) we get

$$\tilde{\sigma}(Tf_0) = \tilde{\sigma}(f_0) + K(\Delta f_0) \quad \text{and} \quad 1-\varepsilon \leq \Delta f < 1 \tag{40}$$

hence

$$co(K(-1,1-O(\varepsilon)) \cup \{2\}) \subset \tilde{\sigma}(g+Tf_0) \subset co(K(-1,1) \cup \{2\}) + K(O(\varepsilon))$$

and hence

$$c_B(g+Tf_o,\ \tfrac{\pi}{2}) \le \sqrt{2} + O(\varepsilon) \quad \text{and}$$

$$c_B(g+Tf_o, \pi + \tfrac{\pi}{2}) \ge 1.7 - O(\varepsilon),$$

so

$$c_B(g+Tf_o,\ \pi+\tfrac{\pi}{2}) - c_B(g+Tf_o,\ \tfrac{\pi}{2}) \ge 0.2 - O(\varepsilon) \tag{41}$$

Put $f_1 = T^{-1}g$. Notice that for any $\emptyset \ne K \subset \mathbb{C}$ and any $c \ge 0$ we have $\rho(K+K(c)) \ge c$ so from (37), and by the definition of g we get that $\Delta g \le 2\varepsilon$ and

$$\tilde{\sigma}(g) = \tilde{\sigma}(f_1) + K(\Delta g) \quad \text{if} \quad 0 \le \Delta g \le 2\varepsilon$$

and

$$\tilde{\sigma}(f_1) = \tilde{\sigma}(g) + K(-\Delta g) \quad \text{if} \quad \Delta g < 0.$$

Hence, since $\| f_o \|_\infty \le \varepsilon$, we get

$$|c_A(f_1+f_o, \pm\tfrac{\pi}{2}) - c_A(f_1, \pm\tfrac{\pi}{2})| \le O(\varepsilon),$$

and

$$|c_A(f_1+f_o,\ \tfrac{\pi}{2}) - c_A(f_1+f_o,\ \pi+\tfrac{\pi}{2})| \le O(\varepsilon). \tag{42}$$

By (36) for any $f \in A$ the difference between $c_A(f,\varphi)$ and $c_B(Tf,\varphi)$ does not depend on φ so this gives the contradiction between (41) and (42).

●

<u>Proof of the second step</u>. Fix $f \in A$ with $\| f \|_\infty = 1$. We can assume as before that there is an $x_o \in \mathrm{Ch}A$ such that $f(x_o) = 1$. Put

$$U_n = \{x \in X : |f(x) - 1| < \tfrac{1}{n^2}\}$$

and let $g_n \in A$ be such that

$$\| g_n \|_\infty \le 1 + \frac{1}{n^2}, \quad g_n(x_o) = 1 \quad \text{and} \quad |g_n(x)| < \frac{1}{n^2}$$

for $x \in X \setminus U_n$. We put $f_n = f + \frac{1}{n} g_n$ and we get

$$1 + \frac{1}{n} \in \tilde{\sigma}(f_n) \subset \mathrm{co}(K(1) \cup \{1 + \tfrac{1}{n}\}) + K(\tfrac{1}{n^2})$$

and hence, by direct computation

$$\rho(\tilde{\sigma}(f_n)) \leq \rho(\tilde{\sigma}(f_n), 1+\tfrac{1}{n}) \leq$$

$$\leq \rho(\mathrm{co}(K(1) \cup \{1+\tfrac{1}{n}\}) + K(\tfrac{1}{n^2}), 1+\tfrac{1}{n}) = \frac{1}{n} + \frac{1}{n^2} < \frac{2}{n} .$$

So we get a sequence $(f_n)_{n=1}^{\infty}$ in A which tends uniformly to f and such that $\rho(\tilde{\sigma}(f_n)) \to 0$.

•

Proof of the third step. For any $K \subset \mathbb{C}$ any any $c \geq 0$ we have $\rho(\mathrm{co}(K) + K(c)) \geq c$ and on the other hand, by (37), if $\Delta f < 0$ then $\tilde{\sigma}(f) = \tilde{\sigma}(Tf) + K(-\Delta f)$ so if $\Delta f < 0$ then $\rho(\tilde{\sigma}(f)) \geq -\Delta f$ and this proves that $\Delta f \geq -\varepsilon$ for all $f \in \mathbf{A}_\varepsilon$.

•

§ 10. Small isomorphisms between natural algebras.

In § 9 we have shown that any isometry between natural Banach algebras which preserves the unit is an algebra isomorphism; that is, we have shown the equivalence between metric and algebraic perturbation in the limit case for $\varepsilon = 0$. In this section we show that it is the only ε for which the equivalence holds, i.e. we construct a natural algebra A and a sequence of onto isomorphisms $T_n : A \to A$ such that $\|T_n\| \|T_n^{-1}\| \xrightarrow[n\to\infty]{} 1$, $T_n(\mathbb{1}) = \mathbb{1}$ but for each $n \in \mathbb{N}$ there is an $f_n \in A$ with $\|f_n\| = 1$ and

$$\| T_n(f_n)T_n(f_n) - T_n(f_n \cdot f_n) \| = 2 .$$

10.1. Example. For any positive integer n we put

$$A_n = \{(a_j)_{j=0}^{\infty} : |a_0| + |a_1| + |a_2| + \sum_{j=3}^{\infty} |a_j| e^{-nj} = \|(a_j)\|_n < \infty\}$$

and define a multiplication on A_n by

$$(a_j) \cdot (b_j) = (c_j) \quad \text{where} \quad c_j = \sum_{i=0}^{j} a_i b_{j-1}.$$

It is easy to check that A_n is a well-defined, commutative, semisimple Banach algebra and that

$$\|(a_j)\|_{\sigma,n} = \sup\{|\sum_{j=0}^{\infty} a_j z^j| : |z| \leq e^{-n}\} \quad \text{for all } (a_j) \in A_n$$

where by $\|\cdot\|_{\sigma,n}$ we denote the spectral norm on A_n. We define a natural norm on A_n by

$$|||(a_j)|||_n = \max\{\|(a_j)\|_{\sigma,n}, \|(a_j)\|_n - |a_0|\}$$

and we define

$$S_n : A_n \to A_n \quad \text{by} \quad S_n((a_j)) = (a_0, a_1, -a_2, a_3, a_4, \ldots,).$$

S_n is a linear isomorphism with $S_n^{-1} = S_n$ and by direct computation we get $|||S_n|||_n \leq 1 + 2e^{-n}$.

We denote by $(e_k)_{k=0}^{\infty}$ the usual Schauder basis of A_n and we have

$$|||e_1|||_n = 1 \quad \text{and} \quad |||S_n(e_1)S_n(e_1) - S_n(e_1e_1)|||_n = 2|||e_2|||_n = 2.$$

To end we let A be a direct sum "in the sense of convergent sequences" of $(A_n)_{n=1}^{\infty}$, i.e.

$$A = \{(a_n)_{n=1}^{\infty} : a_n \in A_n,\ \exists(\lambda_n)_{n=1}^{\infty},\ \sup_n|\lambda_n| < \infty,\ \|a_n - \lambda_n e_0\| \xrightarrow[n\to\infty]{} 0\}$$

and

$$\|(a_n)_{n=1}^{\infty}\| = \sup\{|||a_n||| : n \in \mathbb{N}\}$$

and

$$(a_n)\cdot(b_n) = (a_n \cdot b_n).$$

$(A, \|\cdot\|)$ is a commutative, semisimple, Banach algebra with natural norm. We define T_n from A onto itself by

$$T_n((a_n)) = (a_1, \ldots, a_{n-1}, S_n a_n, a_{n+1}, \ldots).$$

We have $\|T_n\|\cdot\|T_n^{-1}\| = |||S_n|||_n^2 \xrightarrow[n\to\infty]{} 1$ and

$$\|T_n(f_n)T_n(f_n) - T_n(f_n \cdot f_n)\| = 2$$

where $f_n = (0, \ldots, 0, e_1, 0, \ldots)$.

●

Remarks. The first generalization of the Nagasawa theorem to some non-uniform algebras is due to M.Cambern [1] ; in 1965 he proved that if both A and B are equal to $C^1[0,1]$ or to $AC[0,1]$ then any isometry of A onto B is induced by a homeomorphism of the unit interval [0,1]. Here $C^1[0,1]$ is the algebra of complex valued continuously differentiable functions on [0,1] with the norm given by

$$\| f \| = \max_{0 \leq t \leq 1} \{ |f(t)| + |f'(t)| \} \quad \text{for} \quad f \in C^1[0,1]$$

and AC[0,1] is an algebra of complex valued absolutely continuous functions on [0,1] with the norm

$$\| f \| = \|f\|_{\infty} + \|f'\|_1 = \max_{0 \leq t \leq 1} |f(t)| + \int_0^1 |f'(t)| dt \quad \text{for} \quad f \in AC[0,1].$$

In 1971 N.V.Rao and A.K.Roy [1] proved that the same holds for algebras of Lipschitz functions and continuously differentiable functions, both with the following norm

$$\| f \| = \|f\|_{\infty} + \| f' \|.$$

In 1981 M.Cambern and V.Pathak [1] proved the same for $C_0^1(X)$, where X is a closed subset of the real line containing no isolated points.

Recently V.Pathak [1,2] extended the above result to $C^{(n)}[0,1]$ and AC(X), X-closed subset of the real line.

The majority of the results of this chapter is contained in the author's paper [8].

IV. PERTURBATIONS OF OPERATOR ALGEBRAS

§ 11. Introduction.

This chapter is devoted to the study of some operator algebras from the point of view of whether the metric and algebraic perturbations produce the same class of algebras. First we prove that if X and Y are strictly convex Banach spaces and T is an isometry from the algebra $K(X)$ of all compact operators from X into itself, onto $K(Y)$ then $K(X)$ and $K(Y)$ are isomorphic in the category of Banach algebras. Next we show that this is no longer true, in general, for ε-isometries but it remains true if we restrict ourself to uniformly convex spaces.

For this we need some definitions and notation:
For Banach spaces U and V :

- $B(V)$ denotes the closed unit ball of V;
- $E(V)$ denotes the set of extreme points of $B(V)$;
- $U \check{\otimes} V$ denotes the injective tensor product of U and V;
- $L(U,V)$ denotes the Banach space of all continuous linear operators from U into V; if $U = V$ we write $L(U)$ in place of $L(U,U)$;
- $F(U,V)$ denotes the closure in $L(U,V)$ of the algebra of finite dimensional operators, if $U = V$ we write $F(U)$ in place of $F(U,U)$ (notice that if V has the approximation property then $F(U,V)$ is the algebra of all compact operators);
- by $U \simeq V$ we mean that U and V are isometric.

Throughout this chapter we identify $F(U,V)$ with $U^* \check{\otimes} V$ and we frequently view a Banach space V as a subspace of $C(E(V^*))$ or $C(B(V^*))$ where $E(V^*) \subset B(V^*)$ are equipped with the weak $*$ topology. The space $V \check{\otimes} W$ is regarded as a subspace of $C(B(V^*) \times B(W^*))$.

A linear isometry T from $F(U)$ onto $F(V)$ we call canonical if one of the following three possibilities holds:

a) $U \simeq V$ and

$$T = \Phi_0 \otimes \Psi_0$$

where $\Phi_0 \colon U^* \to V^*$ and $\Psi_0 \colon U \to V$ are onto isometries;

b) $U^* \cong V$ and T is of the form

$$T(\Phi) = \Psi_o \circ \Phi^* \circ \Phi_o \qquad \text{for all} \qquad \Phi \in F(U),$$

where $\Psi_o: U^* \to V$ and $\Phi_o: V \to U^*$ are onto isometries;

c) $U \cong V^*$ and T^{-1} if of the form b).

Notice that if there is a canonical isometry from $F(U)$ onto $F(V)$ then $F(U)$ and $F(V)$ or $F(U)$ and $F(V^*)$ are isomorphic in the category of Banach algebras and for finite dimensional Banach spaces U and V any canonical isometry T from $L(U)$ onto $L(V)$ such that $T(Id_U) = Id_V$ is an algebra isomorphism or antiisomorphism.

Notice also that if there exists a canonical onto isometry of the form b) or c) then U and V have to be reflexive, so b) and c) are equivalent.

For a Banach space V we denote by $Mult(V)$ the multiplier algebra of V. $Mult(V)$ consists of all linear and continuous operators $S: V \to V$ such that every extreme functional is an eigenvector of the conjugate operator S^*. By $Z(V)$ we denote the centralizer of V. All fundamental results on multiplier algebra and centralizer are to be found in Behrends [1].

§ 12. Isometries in operator algebras.

12.1. THEOREM. Let X and Y be Banach spaces such that X^{**} and Y^{**} are strictly convex, then any isometry from $F(X)$ onto $F(Y)$ is canonical.

For finite dimensional Banach spaces we get the following stronger result:

12.2. THEOREM. Let X and Y be finite dimensional Banach spaces and assume one of the spaces X, X^*, Y, Y^* is strictly convex. Then any isometry from $L(X)$ onto $L(Y)$ is canonical.

The above theorems are fulfilled as well in the real and the complex case.

We will prove both of the above theorems together.

12.3. Lemma. Let U and V be any linear spaces and assume

$$u_1 \otimes v_1 + u_2 \otimes v_2 = u_3 \otimes v_3$$

where $u_i \in U$, $v_i \in V$, $i = 1,2,3$; then either the vectors u_1, u_2, u_3 or the vectors v_1, v_2, v_3 are proportional.

•

Proof. Let v^* be any linear functional on V. We have

$$u_1 v^*(v_1) + u_2 v^*(v_2) = u_3 v^*(v_3)$$

hence, if $v^*(v_3) \neq 0$ then u_3 is a linear combination of u_1 and u_2; if u_1, u_2, u_3 were not proportional then the coefficients of this linear combination would be uniquely determined and this would mean

$$v^*(v_1) = \text{const } v^*(v_3), \qquad v^*(v_2) = \text{const } v^*(v_3)$$

for any $v^* \in V^*$.

Hence $v_1 \| v_2$ and $v_2 \| v_3$.

•

12.4. Lemma. Let U and V be Banach spaces then

$$E((U \check{\otimes} V)^*) = E(U^*) \check{\otimes} E(V^*).$$

•

Proof. We have

$$U \check{\otimes} V \subseteq \{f \in C(B(U^*) \times B(V^*)) : \quad \forall u^* \in U^* \quad f(u^*, \cdot) \in V \text{ and } \forall v^* \in V^* \quad f(\cdot, v^*) \in U\}.$$

So the lemma is an immediate consequence of the following fact.

Let S be a compact Hausdorff space, let A be a closed subspace of $C(S)$ which separates points of S and let $s \in S$ be such that the functional $\varphi_s \in A^*$: $\varphi_s(f) = f(s)$ has norm one. Then φ_s is an extreme point of the unit ball of A^* if and only if the measure concentrated in $\{s\}$ is the unique probability measure μ on S such that $\int f d\mu = f(s)$ for all $f \in A$.

•

12.5. Lemma. Let $X, \tilde{X}, Y, \tilde{Y}$ be Banach spaces with Y^* being strictly convex and let $T: X \check{\otimes} \tilde{X} \to Y \check{\otimes} \tilde{Y}$ be a linear onto isometry. Then for any $\tilde{y}_o^* \in E(\tilde{Y}^*)$ one of the following possibilities holds:

a) there is an $\tilde{x}_o^* \in E(\tilde{X}^*)$ and a continuous, linear map $\Phi: X \to Y$ such that

$$T^*(y^* \otimes \tilde{y}_o^*) = \Phi^*(y^*) \otimes \tilde{x}_o^* \qquad \text{for all} \qquad y^* \in Y^*;$$

b) there is an $x_o^* \in E(X^*)$ and a continuous, linear map $\Psi: \tilde{X} \to Y$ such that

$$T^*(y^* \otimes \tilde{y}_o^*) = x_o^* \otimes \Psi^*(y^*) \qquad \text{for all} \qquad y^* \in Y^*.$$

Moreover $\Phi^*: Y^* \to X^*$ and $\Psi^*: Y^* \to \tilde{X}^*$ are isometric embeddings.

•

Proof. The map $T^*: (Y \check{\otimes} \tilde{Y})^* \to (X \check{\otimes} \tilde{X})^*$ is an onto isometry so it maps the extreme points of the unit ball onto the extreme points of the unit ball and by Lemma 12.4 we get for any $y_1^*, y_2^*, y_3^* \in E(Y^*)$ there are $x_1^*, x_2^*, x_3^* \in E(X^*)$ and $\tilde{x}_1^*, \tilde{x}_2^*, \tilde{x}_3^* \in E(\tilde{X}^*)$ such that

$$T^*(y_i^* \otimes \tilde{y}_o^*) = x_i^* \otimes \tilde{x}_i^* \qquad \text{for} \quad i = 1,2,3.$$

Assume $y_1^* + y_2^* \neq 0$. The space Y^* being strictly convex it follows that $(y_1^* + y_2^*)/\|y_1^* + y_2^*\|$ is an extreme point of $B(Y^*)$, so

$$x_1^* \otimes \tilde{x}_1^* + x_2^* \otimes \tilde{x}_2^* = T^*((y_1^* + y_2^*) \otimes \tilde{y}_o^*) = \|y_1^* + y_2^*\| x_4^* \otimes \tilde{x}_4^*$$

for some $x_4^* \in E(X^*)$ and $\tilde{x}_4^* \in E(\tilde{X}^*)$. Hence by Lemma 12.3 we have $x_1^* \| x_2^*$ or $\tilde{x}_1^* \| \tilde{x}_2^*$. The same arguments show that $x_1^* \| x_3^*$ or $\tilde{x}_1^* \| \tilde{x}_3^*$ and $x_2^* \| x_3^*$ or $\tilde{x}_2^* \| \tilde{x}_3^*$ and this proves that $x_1^* \| x_2^* \| x_3^*$ or $\tilde{x}_1^* \| \tilde{x}_2^* \| \tilde{x}_3^*$. Hence the strict convexity of Y^* imply now that for any $\tilde{y}_o^* \in E(\tilde{Y}^*)$ one of the following two possibilities takes place:

a. There is an $\tilde{x}_o^* \in E(\tilde{X}^*)$ and a linear, weak * continuous into isometry $\phi: Y^* \to X^*$ such that

$$T^*(y_o^* \otimes \tilde{y}_o^*) = \phi(y^*) \otimes \tilde{x}_o^* \qquad \text{for all} \quad y^* \quad \text{in} \quad Y^*;$$

b. There is an $x_o^* \in E(X^*)$ and a linear, weak * continuous into isometry $\psi: Y^* \to \tilde{X}^*$ such that

$$T^*(y^* \otimes \tilde{y}_o^*) = x_o^* \otimes \psi(y^*) \qquad \text{for all} \quad y^* \quad \text{in} \quad Y^*.$$

To end the proof of the lemma we put $\Phi = \phi^*|_X$ and $\Psi = \psi^*|_X$.
$\bullet$

Notice now that if $X, \tilde{X}, Y, \tilde{Y}$ are of the same, finite dimension then maps Φ and Ψ defined in Lemma 12.5 have to be onto isometries so in this case if Y^* is strictly convex then also one of the spaces X^* or $\tilde{X}^*$ have to be strictly convex. The above shows that if there is an onto isometry from $L(X) \simeq X \check{\otimes} X^* \simeq L(X^*)$ onto $L(Y) \simeq Y \check{\otimes} Y^* \simeq L(Y^*)$ with X, Y being finite dimensional and if X or X^* is strictly convex then also Y or Y^* is strictly convex. So to end, it is sufficient to restrict ourselves to proving Theorem 12.1.

Let X and Y be as in Theorem 12.1 and let T be an onto linear isometry from $F(X) \simeq X \check{\otimes} X^*$ onto $F(Y) \simeq Y \check{\otimes} Y^*$. We can assume $\dim(Y) \geq 2$ since otherwise the theorem is trivial. By Lemma 12.5 for any $y_o^* \in E(Y^*)$ one of the following two possibilities holds:

a. There is an $x_o^* \in E(X^*)$ and a linear, continuous map $\Phi: X^* \to Y^*$ such that

$$T^*(y_o^* \otimes y^{**}) = x_o^* \otimes \Phi^*(y^{**}) \qquad \text{for all} \quad y^{**} \in Y^{**};$$

b. There is an $x_o^{**} \in E(X^{**})$ and a linear, continuous map $\Psi: X \to Y^*$ such that

$$T^*(y_o^* \otimes y^{**}) = \Psi^*(y^{**}) \otimes x_o^{**} \qquad \text{for all} \quad y^{**} \in Y^{**}.$$

By Lemma 12.5 applying to both T and T^{-1} we get that Φ^*, and hence Φ too, are onto isometries. We shall show that Ψ also has to be an onto isometry. Assume to the contrary, let y_o^*, x_o^{**}, Ψ be as in the point b. above, and let $x_1^* \in E(X^*) \setminus \Psi^*(Y^{**})$. Fix

$y^{**} \in E(Y^{**})$. The space X^{**} is strictly convex so X^* is smooth and hence there are numbers $\alpha, \beta \neq 0$ such that $\alpha x_1^* + \beta\Psi^*(y^{**})$ is an extreme point of the unit ball of X^*. By Lemma 12.5 there are $y_i^* \in E(Y^*)$, $y_i^{**} \in E(Y^{**})$, $i = 1,2$ such that

$$(T^{-1})^*((\alpha x_1^* + \beta\Psi^*(y^{**})) \otimes x_o^{**}) = y_1^* \otimes y_1^{**}$$

and also

$$(T^{-1})^*((\alpha x_1^* + \beta\Psi^*(y^{**})) \otimes x_o^{**}) = \alpha(T^{-1})^*(x_1^* \otimes x_o^{**}) + \beta y_o^* \otimes y^{**} =$$

$$= \alpha y_2^* \otimes y_2^{**} + \beta y_o^* \otimes y^{**}.$$

Hence, by Lemma 12.3 we have $y_2^* \| y_o^*$ or $y_2^{**} \| y^{**}$, but $y_2^* \| y_o^*$ contradicts our assumption that $x_1^* \notin \Psi^*(Y^{**})$, so $y_2^{**} \| y^{**}$. Since y^{**} was an arbitrary point of $E(Y^{**})$ we have proven that y_2^{**} is proportional to any point of $E(Y^{**})$ which is absurd since Y is not one dimensional.

We have shown that both Φ and Ψ have to be onto isometries, so T being one to one proves that for all $y_o^* \in E(Y^*)$ the same possibility a. or b. always holds. If the second one holds then X and Y^* and also by symmetry Y and X^* are isometric so X and Y are reflexive and composing T with the isometry $F(Y) \simeq F(Y^*)$ we have situation a. So without loss of generality we can assume that for any $y_o^* \in E(Y^*)$ a. holds.

Let us recapitulate what we have proven:

There is a map $A: E(Y^*) \to E(X^*)$ and a map $B: E(Y^*) \times X^* \to Y^*$ such that

$$T^*(y^* \otimes y^{**}) = A(y^*) \otimes B^*(y^*, y^{**}) \quad \text{for all } y^* \in E(Y^*) \text{ and } y^{**} \in Y^{**} \tag{43}$$

where by $B^*(y^*, \cdot)$ we have denoted the map conjugate to $B(y^*, \cdot)$.

By symmetry there is also a map $\widetilde{A}: E(X^*) \to E(Y^*)$ and a map $\widetilde{B}$: $E(X^*) \times Y^* \to X^*$ such that

$$(T^{-1})^*(x^* \otimes x^{**}) = \widetilde{A}(x^*) \otimes \widetilde{B}^*(x^*, x^{**}) \quad \text{for all } x^* \in E(X^*) \text{ and } x^{**} \in X^{**}. \tag{44}$$

By (43) and (44) the equality $T \circ T^{-1} = \mathrm{Id}$ gives

$$y^* \otimes y^{**} = \tilde{A} \circ A(y^*) \otimes \tilde{B}^*(A(y^*), B^*(y^*, y^{**})) \quad \text{for any } y^* \in E(Y^*) \text{ and } y^{**} \in Y^{**}.$$

Hence

$$\tilde{A} \circ A(y^*) = \alpha(y^*) \cdot y^* \qquad \text{for all } y^* \in E(Y^*), \tag{45}$$

where α is a function defined on $E(Y^*)$ whose values are scalar of modulus one.

12.6. Lemma. For any $y_1^*, y_2^* \in E(Y^*)$ the linear maps $B(y_1^*, \cdot)$ and $B(y_2^*, \cdot)$ are proportional that is, there is a number $\lambda \neq 0$ such that $B(y_1^*, x^*) = \lambda B(y_2^*, x^*)$ for all $x^* \in X^*$.

•

Assume we have the lemma. This shows that multiplying $A(y^*)$ and $B(y^*, \cdot)$ for any $y^* \in E(Y^*)$ by an appropriate scalar of modulus one we can assume without loss of generality that the function B does not depend on the first coordinate that is, there is a map $A_o \colon E(Y^*) \to E(X^*)$ and a map $B_o \colon X^* \to Y^*$ such that

$$T^*(y^* \otimes y^{**}) = A_o(y^*) \otimes B_o^*(y^{**}) \quad \text{for all } y^* \in E(Y^*) \text{ and } y^{**} \in Y^{**}.$$

The above proves that A_o can be extended to a weak $*$ continuous linear isometry from Y^* onto X^* and hence T is a tensor product of $A_o^*|_X$ and B_o. So to end we should prove Lemma 12.6. The proof of the lemma is based on the following statement.

12.7. THEOREM. Let X be a Banach space and assume X is strictly convex then $\dim(\mathrm{Mult}(X)) = 1$.

The above theorem may be interesting in itself and its proof is not very short so before passing to the proof of this theorem we first deduce from it our lemma and so shall end the proof of theorems 12.1 and 12.2. Notice please that if X^{**} is strictly convex then X is also a strictly convex Banach space.

Proof of Lemma 12.6. To prove that two linear operators $\Phi_1 \colon U \to V$ and $\Phi_2 \colon U \to V$ are proportional it is sufficient to show that for any

$u \in U$ the vectors $\Phi_1(u)$ and $\Phi_2(u)$ are proportional, moreover Φ_1 and Φ_2 are proportional if the conjugate operators are so. Hence, assuming to the contrary we get that there is a $y_0^{**} \in E(Y^{**})$ such that $B^*(y_1^*, y_0^{**})$ and $B^*(y_2^*, y_0^{**})$ are not proportional. Let $x_0^{**} \in X^{**}$, $x_0^* \in E(X^*)$, $y_0^* \in E(Y^*)$ are such that

$$B^*(y_1^*, y_0^{**})(x_0^*) = 1, \qquad B^*(y_2^*, y_0^{**})(x_0^*) = 0,$$

$$\widetilde{B}^*(A(y_1^*), x_0^{**})(y_0^*) \neq 0.$$

We define continuous maps $S_1 \colon X \to Y$ and $S_2 \colon Y \to X$ by

$$y^*(S_1(x)) = T^*(y^* \otimes y_0^{**})(x \otimes x_0^*) = A(y^*) \otimes B^*(y^*, y_0^{**})(x \otimes x_0^*) =$$

$$= (A(y^*))(x) \cdot B^*(y^*, y_0^{**})(x_0^*) \quad \text{for all} \quad y^* \in E(Y^*), \quad x \in X;$$

$$x^*(S_2(y)) = (T^{-1})^*(x^* \otimes x_0^{**})(y \otimes y_0^*) =$$

$$= (\widetilde{A}(x^*))(y) \cdot \widetilde{B}^*(x^*, x_0^{**})(y_0^*) \quad \text{for all} \quad x^* \in E(X^*), \quad y \in Y.$$

From (43), (44) and (45) we get that the map $S_1 \circ S_2 \colon Y \to Y$ is of the form

$$y^*(S_1 \circ S_2(y)) = A(y^*)(S_2(y)) \cdot B^*(y^*, y_0^{**})(x_0^*) =$$

$$= \widetilde{A}(A(y^*))(y) \cdot \widetilde{B}^*(A(y^*), x_0^{**})(y_0^*) \cdot B^*(y^*, y_0^{**})(x_0^*) =$$

$$= y^*(y) \cdot \alpha(y^*) \cdot \beta(y^*) \cdot \gamma(y^*) \quad \text{for all} \quad y^* \in E(Y^*), \quad y \in Y,$$

where $\alpha \cdot \beta \cdot \gamma$ is a scalar valued function defined on $E(Y^*)$ such that $\alpha \cdot \beta \cdot \gamma(y_1^*) \neq 0$ and $\alpha \cdot \beta \cdot \gamma(y_2^*) = 0$. Hence $S_1 \circ S_2 \colon Y \to Y$ is a non-trivial multiplier which, by Theorem 12.7, contradicts the assumption Y^{**} is strictly convex.

<u>Proof of Theorem 12.7</u>. Assume that the algebra $\mathrm{Mult}(X)$ is not trivial. Let $x_0 \in X$, $T \in \mathrm{Mult}(X)$ be such that

$$\|x_0\| = 1 = \|T\|, \qquad Tx_0 \neq 0, \qquad x_0 \text{ and } Tx_0 \text{ are not proportional}$$

(T can be taken any element of $\mathrm{Mult}(X)$ not of the form $\lambda \mathrm{Id}_X$ and then an appropriate x_0 exists). For any $S \in \mathrm{Mult}(X)$ by M_S we denote a continuous function on $M = \overline{E(X^*)} \setminus \{0\}$ such that

$$S^*(x^*) = M_S(x^*) \cdot x^* \qquad \text{for all} \quad x^* \in M, \tag{46}$$

where the closure is taken in the weak $*$ topology. Put

$$G = M_T(M),$$

$$f_o(z) = \sup \{|x^*(x_o)| : x^* \in M_T^{-1}(z)\} \qquad \text{for } z \in G$$

$$r_o = \inf \{r \geq 0: \ r(2 - \operatorname{Re} z) \geq f_o(z) \qquad \text{for any } z \in G\}.$$

We have

$$G \subset \{z \in \mathbb{C}: \ |z| \leq 1\},$$

$$0 \leq f_o(z) \leq 1 \qquad \text{for } z \in G,$$

$$0 < r_o < \infty.$$

Let $z_o \in G$ be such that

$$r_o(2 - \operatorname{Re} z_o) = f_o(z_o) \tag{47}$$

(such a z_o exists provided f_o is upper semicontinuous and for any sequence $z_n \in G$ such that $z_n \mapsto z_o \in \overline{G} \setminus G$ we have $f_o(z_n) \to 0$). Put

$$\kappa(z) = k(z - w_o)^2 \qquad \text{for } z \in \mathbb{C}$$

where k and w_o are such that the plane in $\mathbb{C} \times \mathbb{R} \simeq \mathbb{R}^3$ given by $z \mapsto r_o(2 - \operatorname{Re} z)$ is tangential to the surface $z \mapsto |\kappa(z)|$ in the point $(z_o, r_o(2 - \operatorname{Re} z_o))$.

Since the function $z \mapsto |\kappa(z)|$ is real analytic there is a positive integer p such that

$$\liminf_{z \to z_o} \frac{|\kappa(z)| - r_o(2 - \operatorname{Re} z)}{|z - z_o|^p} = a > 0 \tag{48}$$

(by a direct computation we get $p = 2$ and $a = |k|$); hence for any sufficiently small complex number ε we have

$$|\kappa(z) + \varepsilon(z - z_o)^p| \geq r_o(2 - \operatorname{Re} z) \qquad \text{for } z \in G. \tag{49}$$

We have assumed that x_o and $Tx_o \neq 0$ are not proportional so there is a $z_1 \in G$ such that $z_1 \neq z_2$ and $f_o(z_1) = \sup \{|x^*(x_o)| : x^* \in M_T^{-1}(z_1)\} > 0$. Put

$$\kappa_\varepsilon(z) = \kappa(z) + \varepsilon(z - z_o)^p \qquad \text{for} \quad z \in G$$

and assume ε is such that $\kappa_\varepsilon(z_1) \neq \kappa(z_1)$ and (49) is satisfied. Put

$$T_1 = \frac{1}{\kappa}(T), \qquad T_2 = \frac{1}{\kappa_\varepsilon}(T), \qquad x_1 = T_1(x_o), \qquad x_2 = T_2(x_o).$$

We have

$$M_{T_1} = \frac{1}{\kappa}(M_T) \qquad \text{and} \qquad M_{T_2} = \frac{1}{\kappa_\varepsilon}(M_T)$$

so $x_1 \neq x_2$. We get

$$\|x_1\| = \sup \{|x^*(x_1)| : x^* \in M\} = \sup_{z \in G} \sup_{x^* \in M_T^{-1}(z)} \left| \frac{x^*(x_o)}{\kappa(z)} \right| =$$

$$= \sup \{ \frac{f_o(z)}{|\kappa(z)|} : z \in G\} = 1.$$

In the same manner we get $\|x_2\| = 1$. Moreover

$$\|x_1 + x_2\| \geq |(\frac{1}{\kappa} + \frac{1}{\kappa_\varepsilon})(z_o)f_o(z_o)| \geq 2,$$

so we have the following situation

$$\|x_1\| = \|x_2\| = \|\frac{x_1+x_2}{2}\| = 1 \qquad \text{and} \qquad x_1 \neq x_2$$

and this contradicts the assumption X is strictly convex.

§ 13. Small isomorphisms in operator algebras.

13.1. Example. Let $X_o = (\mathbb{R}^2, \|\cdot\|_\infty)$ be a two dimensional real Banach space with the norm defined as a supremum of the modulus of coordinates. For any $\varepsilon > 0$ fix a strictly convex, smooth two dimensional real Banach space $X_\varepsilon = (\mathbb{R}^2, \|\cdot\|_\varepsilon)$ such that the Banach-Mazur distance between X_ε and X_o is less than $1 + \varepsilon$. Elements of $B(X_\varepsilon)$ we can identify with 2×2 matrices. For $\varepsilon \geq 0$ we define $T_\varepsilon : B(X_\varepsilon) \to B(X_\varepsilon)$ by

$$T_\varepsilon \left(\begin{bmatrix} a & b \\ c & d \end{bmatrix} \right) = \begin{bmatrix} \frac{a+b+c+d}{2} & \frac{a+b-c-d}{2} \\ \frac{a+c-b-d}{2} & \frac{a+d-b-c}{2} \end{bmatrix}$$

We have $\|T_o\| = \|T_o^{-1}\| = 1$ so $\lim_{\varepsilon\to 0} \|T_\varepsilon\|\,\|T_\varepsilon^{-1}\| = 1$ and $T_\varepsilon(Id_{X_\varepsilon}) = Id_{X_\varepsilon}$. But

$$\left\| T_\varepsilon\left(\begin{bmatrix}1&0\\0&0\end{bmatrix}\right) \cdot T_\varepsilon\left(\begin{bmatrix}0&1\\1&0\end{bmatrix}\right) - T_\varepsilon\left(\begin{bmatrix}1&0\\0&0\end{bmatrix} \circ \begin{bmatrix}0&1\\1&0\end{bmatrix}\right)\right\|_\varepsilon =$$

$$= \left\|\begin{bmatrix}0&-1\\1&0\end{bmatrix}\right\|_\varepsilon \xrightarrow[\varepsilon\to 0]{} \left\|\begin{bmatrix}0&-1\\1&0\end{bmatrix}\right\|_\infty = 1$$

so T is not $O(\varepsilon)$-almost multiplicative.

•

The above example shows that Theorem 12.1 is no longer true for ε-isometries. We shall now prove that it remains true for uniformly convex spaces.

<u>13.2. THEOREM</u>. Let $X,\tilde{X},Y,\tilde{Y}$ be Banach spaces with uniformly convex duals. Then there is an $\varepsilon_o > 0$ such that for any $\varepsilon \le \varepsilon_o$ and any linear isomorphism T from $X \check{\otimes} \tilde{X}$ onto $Y \check{\otimes} \tilde{Y}$ with $\|T\|\cdot\|T^{-1}\| \le 1+\varepsilon$ there are linear isomorphisms $\Phi: X \to Y$ and $\Psi: \tilde{X} \to \tilde{Y}$ or $\Phi: X \to \tilde{Y}$ and $\Psi: \tilde{X} \to Y$ with $\|\Phi\|\,\|\Phi^{-1}\| \le 1 + c(\varepsilon)$ and $\|\Psi\|\,\|\Psi^{-1}\| \le 1 + c(\varepsilon)$ such that $\|T - \Phi \otimes \Psi\| \le c(\varepsilon)$. The constant ε_o and the function c depend only on the modulus of convexity of the considered Banach spaces and $\lim_{\varepsilon\to 0^+} c(\varepsilon) = 0$.

<u>13.3. Corollary</u>. Let X,Y be Banach spaces with the approximation property and such that X,X^*,Y and Y^* are uniformly convex. Then there is an $\varepsilon_o > 0$ such that if the Banach-Mazur distance between $K(X)$ and $K(Y)$ is less than $1+\varepsilon_o$ then $K(X)$ and $K(Y)$ or $K(X)$ and $K(Y^*)$ are isomorphic in the category of Banach algebras. The constant ε_o depends only on the modulus of convexity of X,X^*,Y and Y^*.

13.4. Corollary. Let X,Y be finite dimensional Banach spaces such that X,X^*,Y and Y^* are strictly convex. Then there is an $\varepsilon_o > 0$ such that for any $\varepsilon \leq \varepsilon_o$ and any linear map T from $L(X)$ onto $L(Y)$ with $\|T\| \|T^{-1}\| \leq 1+\varepsilon$ and $T(Id_X) = Id_Y$ there is an algebra isomorphism or an algebra antiisomorphism $\tilde{T}$ from $L(X)$ onto $L(Y)$ such that $\|T-\tilde{T}\| \leq c'(\varepsilon)$; where ε_o and c' depend only of the modulus of convexity of X,X^*,Y,Y^*.

Proof of the theorem. We assume, without loss of generality, that $\|T\| \leq 1+\varepsilon$ and $\|T^{-1}\| \leq 1+\varepsilon$.

At various points of the proof we shall use the inequalities involving ε which are valid only if ε is sufficiently small, in those cases we will merely assume that ε is near 0 and this assumption gives rise to the constant ε_o.

13.5. Lemma. Let U and V be normed, linear spaces, let δ be positive and assume that

$$\|u_1 \otimes v_1 + u_2 \otimes v_2 + u_3 \otimes v_3\| \leq \delta \tag{50}$$

where $u_1,u_2,u_3 \in U$, $v_1,v_2,v_3 \in V$ and $\|u_1\| = \|u_2\| = 1 = \|v_1\| = \|v_2\| = \|v_3\|$. Then there is a number λ of modulus one such that $\|u_1 - \lambda u_2\| \leq 3\sqrt{\delta}$ or $\|v_1 - \lambda v_2\| \leq 3\sqrt{\delta}$.

Proof of the lemma. If $\inf_{|\lambda|=1} \|\lambda v_i - v_3\| \leq \frac{3}{2}\sqrt{\delta}$ for both $i=1$ and 2 then we get $\|v_1 - \lambda v_2\| \leq 3\sqrt{\delta}$ for some λ of modulus one, so we can assume that

$$\inf_{|\lambda|=1} \|\lambda v_1 - v_3\| > \frac{3}{2}\sqrt{\delta}. \tag{51}$$

Assume there is an $\alpha \in \mathbb{C}$ with $|\alpha v_3 - v_1| \leq \frac{3}{4}\sqrt{\delta}$. We get

$$1 + \frac{3}{4}\sqrt{\delta} \geq |\alpha| \geq 1 - \frac{3}{4}\sqrt{\delta}$$

and hence, by (51)

$$\left\|\frac{\alpha}{|\alpha|}v_3 - v_1\right\| \leq \left|\frac{\alpha}{|\alpha|} - \alpha\right| + \|\alpha v_3 - v_1\| \leq \left|\frac{\alpha(1-|\alpha|)}{|\alpha|}\right| + \frac{3}{4}\sqrt{\delta} \leq \frac{3}{2}\sqrt{\delta}.$$

The above contradicts (51) and we get

$$\inf_{\alpha\in\mathbb{C}} \|\alpha v_3 - v_1\| > \frac{3}{4}\sqrt{\delta}. \tag{52}$$

We define a functional v^* on $\mathrm{span}(v_1,v_3)$ by

$$v^*(\alpha v_1 + \beta v_3) = \frac{3}{4}\sqrt{\delta}\alpha.$$

From (52) we have $\|v^*\| \leq 1$. Let $\tilde{v}^*$ be a norm preserving extension of v^* from $\mathrm{span}(v_1,v_3)$ to V. From (50) we get

$$\|u_1\tilde{v}^*(v_1) + u_2\tilde{v}^*(v_2)\| \leq \delta,$$

so

$$\left\|u_1 + u_2\frac{\tilde{v}^*(v_2)}{\tilde{v}^*(v_1)}\right\| \leq \frac{4}{3}\sqrt{\delta}.$$

Hence, in the same manner as before we get

$$\left\|u_1 + u_2\frac{\tilde{v}^*(v_2)}{\tilde{v}^*(v_1)}\frac{|\tilde{v}^*(v_1)|}{|\tilde{v}^*(v_2)|}\right\| \leq 2\cdot\frac{4}{3}\sqrt{\delta} < 3\sqrt{\delta}.$$

•

For the next lemmas we need the following observations. The first one is easy to check by direct computation.

<u>13.6. Proposition</u>. Let V be a Banach space with uniformly convex dual and let $v \in V$, $\|v\| = 1$ then

$$\mathrm{diam}\ \{v^* \in B(V^*): \mathrm{Re}(v^*(v)) \geq 1 - \delta\} \leq \delta^*_{V^*}(2\delta).$$

<u>13.7. Proposition</u>. Let V,U be Banach spaces with uniformly convex duals and let $v \in V$, $u \in U$, $\|v\| = 1 = \|u\|$ then

$$\operatorname{diam}\{v^* \otimes u^* \in B(V^*) \check{\otimes} B(U^*) : \operatorname{Re}(v^* \otimes u^*(v \otimes u)) \geq 1 - \delta\} \leq$$
$$\leq \delta^*_{V^*}(2\delta) + \delta^*_{U^*}(2\delta).$$

Proof. Fix $v_i^* \otimes u_i^* \in B(V^*) \check{\otimes} B(U^*)$ such that

$$\operatorname{Re}(v_i^* \otimes u_i^*)(v \otimes u) \geq 1 - \delta \qquad \text{for } i = 1,2.$$

Let α_i, $i = 1,2$ be complex numbers of modulus one such that $\alpha_i v_i^*(v) \in \mathbb{R}^+$. By our assumption we get

$$\alpha_i v_i^*(v) \geq 1 - \delta \quad \text{and} \quad \operatorname{Re} \frac{1}{\alpha_i} u_i^*(u) \geq 1 - \delta \quad \text{for } i = 1,2.$$

Hence by Proposition 13.6 we get

$$\|\alpha_1 v_1^* - \alpha_2 v_2^*\| \leq \delta^*_{V^*}(2\delta) \quad \text{and} \quad \left\|\frac{1}{\alpha_1} u_1^* - \frac{1}{\alpha_2} u_2^*\right\| \leq \delta^*_{U^*}(2\delta)$$

so

$$\|v_1^* \otimes u_1^* - v_2^* \otimes u_2^*\| \leq \left\|\alpha_1 v_1^* \otimes \frac{1}{\alpha_1} u_1^* - \alpha_2 v_2^* \otimes \frac{1}{\alpha_1} u_1^*\right\| +$$
$$+ \left\|\alpha_2 v_2^* \otimes \frac{1}{\alpha_1} u_1^* - \alpha_2 v_2^* \otimes \frac{1}{\alpha_2} u_2^*\right\| \leq \delta^*_{V^*}(2\delta) + \delta^*_{U^*}(2\delta).$$

●

13.8. Proposition. Let S be a compact Hausdorff space, let A be a closed subspace of $C(S)$ and let F be a norm one functional on A. We denote by S_o the subset of S consisting of all points s from S such that the norm of functional $A \ni f \mapsto f(s)$ is equal to one. Assume that for any $s \in S$ and any number λ of modulus one there is exactly one $s_\lambda \in S$ such that $f(s) = \lambda f(s_\lambda)$ for all f in A. Then there is a probability measure μ on S which is norm preserving extension of F from A to $C(S)$. Furthermore for any such μ we have

$$\mu(S_o) = \mu(S) = 1.$$

Proof. Let ν be a norm one extension of F from A to $C(S)$. Denote by K_r the subset of S consisting of all points $s \in S$ such that the norm of functional $A \ni f \mapsto f(s)$ is not greater than r.

For any $f \in A$ with $\|f\| = 1$ we have

$$|F(f)| = \left|\int_S f d\nu\right| \le \int_{K_r} |f| d|\nu| + \int_{S \setminus K_r} |f| d|\nu| \le$$

$$= \sup\{|f(s)| : s \in K_r\} \cdot |\nu|(K_r) + |\nu|(S \setminus K_r) \le 1 - |\nu|(K_r) \cdot (1-r).$$

Hence $|\nu|(K_r) = 0$ for any $r < 1$.

Put $h = \frac{d\nu}{d|\nu|}$. We can assume $|h| \equiv 1$ on S. By our assumption there is a map $\varphi: S \to S$ such that

$$h(s)f(s) = f \circ \varphi(s) \qquad \text{for } f \in A, \ s \in S.$$

If h is continuous then the corresponding function φ defined by the above equality is also continuous. Hence it is standard to prove that if h is a Borel function then φ is also Borel. To end the proof we define μ by

$$\mu(K) = |\nu|(\varphi^{-1}(K)) \qquad \text{for any Borel subset } K \text{ of } S.$$

●

13.9. Lemma. Let $X, \tilde{X}, Y, \tilde{Y}$ be Banach spaces with uniformly convex duals and let T be a linear isomorphism from $X \check{\otimes} \tilde{X}$ onto $Y \check{\otimes} \tilde{Y}$ with $\|T\| \le 1 + \varepsilon$, $\|T^{-1}\| \le 1 + \varepsilon$. Then for any $y^* \in E(Y^*)$, $\tilde{y}^* \in E(\tilde{Y}^*)$ there are $x^* \in E(X^*)$, $\tilde{x}^* \in E(\tilde{X}^*)$ such that

$$\|T^*(y^* \otimes \tilde{y}^*) - x^* \otimes \tilde{x}^*\| \le \alpha(\varepsilon);$$

where $\alpha(\varepsilon) \xrightarrow[\varepsilon \to 0]{} 0$ and the function α depends only on the modulus of convexity of $X^*, \tilde{X}^*, Y^*, \tilde{Y}^*$.

●

Proof of the lemma. Fix $y_o^* \in E(Y^*)$, $\tilde{y}_o^* \in E(\tilde{Y}^*)$ and let μ be a measure on $B(X^*) \times B(\tilde{X}^*)$ which is a norm preserving extension of the functional $T^*(y_o^* \otimes \tilde{y}_o^*)$ from $X \check{\otimes} \tilde{X}$ to $C(B(X^*) \times B(\tilde{X}^*))$. By Proposition 13.8 we can assume that μ is positive and we have

$$\|\mu\| = \mu(B(X^*) \times B(\tilde{X}^*)) = \mu(E(X^*) \times E(\tilde{X}^*))$$

and

$$1-\varepsilon \leq \|\mu\| \leq 1+\varepsilon .$$

Spaces Y and $\tilde{Y}$ are reflexive so there are $y_o \in B(Y)$, $\tilde{y}_o \in B(\tilde{Y})$ such that $y_o^*(y_o) = 1 = \tilde{y}_o^*(\tilde{y}_o)$.

Put

$$S = \{(x^*,\tilde{x}^*) \in E(X^*) \times E(\tilde{X}^*) : \operatorname{Re}(T^{-1}(y_o \otimes \tilde{y}_o))(x^* \otimes \tilde{x}^*) \geq 1-\sqrt{\varepsilon}\}.$$

We have $\|\mu\| \leq 1+\varepsilon$, $\|T^{-1}(y_o \otimes \tilde{y}_o)\| \leq 1+\varepsilon$ and $\int T^{-1}(y_o \otimes \tilde{y}_o)\,d\mu = 1$ so, by direct calculation

$$\mu(E(X^*) \times E(\tilde{X}^*) \setminus S) \leq 2\sqrt{\varepsilon}. \tag{53}$$

We shall show that

$$\operatorname{diam}(\{(x^* \otimes \tilde{x}^*) : (x^*,\tilde{x}^*) \in S\}) \leq \alpha'(\varepsilon) \tag{54}$$

where $\alpha'(\varepsilon) \xrightarrow[\varepsilon\to 0]{} 0$, and α' depends only on the modulus of convexity of $X^*,\tilde{X}^*,Y^*,\tilde{Y}^*$.

For this purpose let $(x_i^*,\tilde{x}_i^*) \in S$ for $i = 1,2$. Spaces X and $\tilde{Y}$ are reflexive so there are $x_i \in B(X)$, $\tilde{x}_i \in B(\tilde{X})$ such that $x_i^*(x_i) = 1 = \tilde{x}_i^*(\tilde{x}_i)$ for $i = 1,2$. We have

$$\|x_i \otimes \tilde{x}_i + T^{-1}(y_o \otimes \tilde{y}_o)\| \geq |((x_i \otimes \tilde{x}_i) + T^{-1}(y_o \otimes \tilde{y}_o))(x_i^* \otimes \tilde{x}_i^*)| \geq 2-\sqrt{\varepsilon}$$

hence

$$\|T(x_i \otimes \tilde{x}_i) + y_o \otimes \tilde{y}_o\| \geq 2 - 2\sqrt{\varepsilon} \qquad \text{for} \qquad i = 1,2.$$

Let $y_i^* \otimes \tilde{y}_i^* \in E(Y^*) \check{\otimes} E(\tilde{Y}^*)$ be such that

$$\operatorname{Re}(y_i^* \otimes \tilde{y}_i^*(T(x_i \otimes \tilde{x}_i) + y_o \otimes \tilde{y}_o)) \geq 2 - 2\sqrt{\varepsilon}.$$

Hence

$$\operatorname{Re}(y_i^* \otimes \tilde{y}_i^*(T(x_i \otimes \tilde{x}_i))) \geq 1 - 2\sqrt{\varepsilon},$$

$$\operatorname{Re}(y_i^* \otimes \tilde{y}_i^*(y_o \otimes \tilde{y}_o)) \geq 1 - 3\sqrt{\varepsilon}.$$

By Proposition 13.7 we get

$$\|y_o^* \otimes \tilde{y}_o^* - y_i^* \otimes \tilde{y}_i^*\| \leq \delta_{Y^*}^*(6\sqrt{\varepsilon}) + \delta_{\tilde{Y}^*}^*(6\sqrt{\varepsilon})$$

which in view of previous inequalities leads to

$$\operatorname{Re}(T(x_i \otimes \tilde{x}_i))(y_o^* \otimes \tilde{y}_o^*) \geq 1 - 3\sqrt{\varepsilon} - 3(1+\varepsilon)[\delta^*_{Y^*}(6\sqrt{\varepsilon}) + \delta^*_{\tilde{Y}^*}(6\sqrt{\varepsilon})] = \gamma(\varepsilon)$$

for $i = 1,2,$

so

$$\|x_1 \otimes \tilde{x}_1 + x_2 \otimes \tilde{x}_2\| \geq 2\gamma(\varepsilon).$$

Hence there is an $x^* \otimes \tilde{x}^* \in E(X^*) \check{\otimes} E(\tilde{X}^*)$ such that

$$\operatorname{Re}(x_i \otimes \tilde{x}_i(x^* \otimes \tilde{x}^*)) \geq 2\gamma(\varepsilon) - 1 \qquad \text{for both} \quad i = 1 \quad \text{and} \quad 2.$$

By Proposition 13.7

$$\|x_1^* \otimes \tilde{x}_1^* - x_2^* \otimes \tilde{x}_2^*\| \leq \delta^*_{X^*}(4 - 4\gamma(\varepsilon)) + \delta^*_{\tilde{X}^*}(4 - 4\gamma(\varepsilon)) = \alpha'(\varepsilon).$$

Fix $(x_o^*, \tilde{x}_o^*) \in S$. To end the proof we observe that for any $f \in X \check{\otimes} \tilde{X}$ with $\|f\| \leq 1$, it follows from (53) and (54) that

$$|f(x_o^* \otimes \tilde{x}_o^*) - T^*(y_o^* \otimes \tilde{y}_o^*)(f)| = |f(x_o^* \otimes \tilde{x}_o^*) - \int f \, d\mu| \leq$$

$$\leq 4\sqrt{\varepsilon} + \int_S |f - f(x_o^* \otimes \tilde{x}_o^*)| \, d\mu + |1 - \mu(S)| \leq$$

$$\leq 4\sqrt{\varepsilon} + \alpha'(\varepsilon)(1+\varepsilon) + 4\sqrt{\varepsilon} = \alpha(\varepsilon).$$

•

<u>13.10. Lemma</u>. Let $X, \tilde{X}, Y, \tilde{Y}, T, \varepsilon, \alpha$ be as in Lemma 13.9. Assume $y_o^* \in E(Y^*)$, $\tilde{y}_1^*, \tilde{y}_2^*, \tilde{y}_3^* \in E(\tilde{Y}^*)$, $x_1^*, x_2^*, x_3^* \in E(X^*)$, $\tilde{x}_1^*, \tilde{x}_2^*, \tilde{x}_3^* \in E(\tilde{X}^*)$ are such that

$$\|T^*(y_o^* \otimes \tilde{y}_i^*) - x_1^* \otimes \tilde{x}_i^*\| \leq \alpha(\varepsilon) \qquad \text{for} \quad i = 1,2,3,$$

then there are numbers $\lambda_{i,j}$ for $i,j = 1,2,3$ of modulus one such that

$$\|x_i^* - \lambda_{i,j} x_j^*\| \leq \beta(\varepsilon) \qquad \text{for} \quad i,j = 1,2,3$$

or

$$\|\tilde{x}_i^* - \lambda_{i,j} \tilde{x}_j^*\| \leq \beta(\varepsilon) \qquad \text{for} \quad i,j = 1,2,3$$

where $\beta(\varepsilon) = 24\sqrt{\alpha(\varepsilon)}$.

•

<u>Proof of the lemma</u>. Since $\tilde{Y}^*$ is uniformly convex, by Lemma 13.9, there are $x_4^* \in E(X^*)$ and $\tilde{x}_4^* \in E(\tilde{X}^*)$ such that

$$\|T(y_o^* \otimes (\tilde{y}_1^* + \tilde{y}_2^*)) - kx_4^* \otimes \tilde{x}_4^*\| \leq k\alpha(\varepsilon),$$

where $k = \|\tilde{y}_1^* + \tilde{y}_2^*\| \leq 2$. Hence

$$\|x_1^* \otimes \tilde{x}_1^* + x_2^* \otimes \tilde{x}_2^* - kx_4^* \otimes \tilde{x}_4^*\| \leq (k\alpha(\varepsilon) + 2\alpha(\varepsilon)) \leq 4\alpha(\varepsilon),$$

and by Lemma 13.5 we have

$$\|x_1^* - \lambda x_2^*\| \leq 12\sqrt{\alpha(\varepsilon)}$$

or

$$\|\tilde{x}_1^* - \lambda \tilde{x}_2^*\| \leq 12\sqrt{\alpha(\varepsilon)} \quad \text{for some } \lambda \text{ of modulus one.}$$

Considering successively the pairs of indices (1,2), (2,3) and (1,3) we get the assertion of the lemma.

From Lemmas 13.9 and 13.10 we get that for any $y_o^* \in E(Y^*)$ we have exactly two possibilities:

a) there is an $x_o^* \in E(X^*)$ and a function $\phi: E(\tilde{Y}^*) \to E(\tilde{X}^*)$ such that

$$\|T^*(y_o^* \otimes \tilde{y}^*) - x_o^* \otimes \phi(\tilde{y}^*)\| \leq \alpha(\varepsilon) + \beta(\varepsilon) = \gamma(\varepsilon) \quad \text{for all } \tilde{y}^* \in E(\tilde{Y}^*) \tag{55}$$

or

b) there is an $\tilde{x}_o^* \in E(\tilde{X}^*)$ and a function $\psi: E(\tilde{Y}^*) \to E(X^*)$ such that

$$\|T^*(y_o^* \otimes \tilde{y}^*) - \psi(\tilde{y}^*) \otimes \tilde{x}_o^*\| \leq \gamma(\varepsilon) \quad \text{for all } \tilde{y} \in E(\tilde{Y}^*). \tag{56}$$

By the same arguments applied to the map T^{-1} in place of T, we get by symmetry (replacing the space X by $\tilde{X}$ and Y by $\tilde{Y}$) and by Lemma 13.10 that

$$\begin{aligned} &\sup\{\inf\{\|\phi(\tilde{y}^*) - \tilde{x}^*\|: \tilde{y}^* \in E(\tilde{Y}^*)\}: \tilde{x}^* \in E(\tilde{X}^*)\} \leq \gamma(\varepsilon) \\ &\sup\{\inf\{\|\psi(\tilde{y}^*) - x^*\|: \tilde{y}^* \in E(\tilde{Y}^*)\}: x^* \in E(X^*)\} \leq \gamma(\varepsilon). \end{aligned} \tag{57}$$

For any $y_o^* \in E(Y^*)$ we define, depending on which of the above possibilities takes place, a function

$$\Phi: \tilde{X} \to \tilde{Y} \qquad \text{or} \qquad \Psi: X \to \tilde{Y}$$

as follows:

a) fix $x_o \in B(X)$ such that $x_o^*(x_o) = 1$ and define Φ by

$$\tilde{y}^*(\Phi(\tilde{x})) = y_o^* \otimes \tilde{y}^*(T(x_o \otimes \tilde{x})) \qquad \text{for} \quad \tilde{y}^* \in \tilde{Y}^*, \quad \tilde{x} \in \tilde{X};$$

b) fix $\tilde{x}_o \in B(\tilde{X})$ such that $\tilde{x}_o^*(\tilde{x}_o) = 1$ and define Ψ by

$$\tilde{y}^*(\Psi(x)) = y_o^* \otimes \tilde{y}^*(T(x \otimes \tilde{x}_o)) \qquad \text{for} \quad \tilde{y}^* \in \tilde{Y}^*, \quad x \in X.$$

The above definitions may depend on the choice of x_o $(\tilde{x}_o)$ and we assume that we have fixed some Φ (Ψ) as above, for any $y_o^* \in$ $\in E(Y^*)$. We have

$$\|\Phi\| \le 1 + \varepsilon, \qquad \|\Psi\| \le 1 + \varepsilon,$$

and

$$|\tilde{y}^*(\Phi(\tilde{x})) - \phi(\tilde{y}^*)(\tilde{x})| \le \gamma(\varepsilon)\|\tilde{x}\| \qquad \text{for all} \quad \tilde{y}^* \in E(\tilde{Y}^*), \quad \tilde{x} \in \tilde{X},$$

$$|\tilde{y}^*(\Psi(x)) - \psi(\tilde{y}^*)(x)| \le \gamma(\varepsilon)\|x\| \qquad \text{for all} \quad \tilde{y}^* \in E(\tilde{Y}^*), \quad x \in X,$$

so from (57) we enter that Φ and Ψ are one to one, onto isometries with

$$\|\Phi^{-1}\| \le 1 + \gamma(\varepsilon), \qquad \|\Psi^{-1}\| \le 1 + \gamma(\varepsilon),$$

and

$$\left.\begin{aligned} \|\Phi^*(\tilde{y}^*) - \phi(\tilde{y}^*)\| &\le \gamma(\varepsilon) \\ \|\Psi^*(\tilde{y}^*) - \psi(\tilde{y}^*)\| &\le \gamma(\varepsilon) \end{aligned}\right\} \quad \text{for all} \quad \tilde{y}^* \in E(\tilde{Y}^*).$$

To end the proof we show that for all $y_o^* \in E(Y^*)$ one of the two possibilities a) or b) takes place and the map assigning to $y_o^* \in E(Y^*)$ a $\Phi \in L(\tilde{X},\tilde{Y})$ $(\Psi \in L(X,\tilde{Y}))$ is "ε-almost" constant.

To this end assume that $y_1^*, y_2^* \in E(Y^*)$, $x_1^* \in E(X^*)$, $\tilde{x}_2^* \in E(\tilde{X}^*)$, $\Phi_1 \in L(\tilde{X},\tilde{Y})$, $\Psi_2 \in L(X,\tilde{Y})$ are such that

$$\left.\begin{aligned} &\|T^*(y_1^* \otimes \tilde{y}^*) - x_1^* \otimes \Phi_1^*(\tilde{y}^*)\| \le 2\gamma(\varepsilon) && (58) \\ \text{and} \quad &\|T^*(y_2^* \otimes \tilde{y}^*) - \Psi_2^*(\tilde{y}^*) \otimes \tilde{x}_2^*\| \le 2\gamma(\varepsilon) && (59) \end{aligned}\right\} \quad \text{for all} \quad \tilde{y}^* \in E(\tilde{Y}^*).$$

Since $\|(\Phi_1^*)^{-1}\| \le 1 + \gamma(\varepsilon)$, $\|(\Psi_2^*)^{-1}\| \le 1 + \gamma(\varepsilon)$ there are $\tilde{y}_1^*, \tilde{y}_2^* \in E(\tilde{Y}^*)$ such that $\|\Phi_1^*(\tilde{y}_1^*) - \tilde{x}_2^*\| \le \gamma(\varepsilon)$, $\|\Psi_2^*(\tilde{y}_2^*) - x_1^*\| \le \gamma(\varepsilon)$;

so we get

$$\|x_1^* \otimes \tilde{x}_2^* - T^*(y_i^* \otimes \tilde{y}_i^*)\| \leq (1+\varepsilon)\gamma(\varepsilon) + 2\gamma(\varepsilon) \quad \text{for} \quad i = 1,2$$

and hence

$$\|y_1^* \otimes \tilde{y}_1^* - y_2^* \otimes \tilde{y}_2^*\| \leq 2(1+\varepsilon)(3+\varepsilon)\gamma(\varepsilon) \leq 7\gamma(\varepsilon)$$

leading to the inequality

$$\|y_1^* - y_2^*\| \leq 7\gamma(\varepsilon)$$

which contradicts (58) and (59).

Thus without loss of generality we can assume that it is the first possibility that always holds.

Fix $y_o^* \in E(Y^*)$ and $\tilde{y}_o^* \in E(\tilde{Y}^*)$. There is an $x_o^* \in E(X^*)$ and $\Phi_o \in L(\tilde{X},\tilde{Y})$ with $\|\Phi_o\|\,\|\Phi_o^{-1}\| \leq (1+\varepsilon)(1+\gamma(\varepsilon))$ such that

$$\|T^*(y_o^* \otimes \tilde{y}^*) - x_o^* \otimes \Phi_o^*(\tilde{y}^*)\| \leq 2\gamma(\varepsilon) \quad \text{for all} \quad \tilde{y}^* \in E(\tilde{Y}^*), \tag{60}$$

by symmetry, there is an $\tilde{x}_o^* \in E(\tilde{X}^*)$ and $\Psi_o \in L(X,Y)$ with $\|\Psi_o\|\,\|\Psi_o^{-1}\| \leq$ $\leq (1+\varepsilon)(1+\gamma(\varepsilon))$ such that

$$\|T^*(y^* \otimes \tilde{y}_o^*) - \Psi_o^*(y^*) \otimes \tilde{x}_o^*\| \leq 2\gamma(\varepsilon) \quad \text{for all} \quad y^* \in E(Y^*). \tag{61}$$

Moreover replacing $2\gamma(\varepsilon)$ in (60) and (61) by $4\gamma(\varepsilon)$ we can assume $\tilde{x}_o^* = \Phi_o^*(\tilde{y}_o^*)$ and $x_o^* = \Psi_o^*(y_o^*)$.

Let us compose T with $\Phi^{-1} \otimes \Psi^{-1}$. To end the proof it is enough to show the following lemma:

<u>13.8. Lemma</u>. Let $X,\tilde{X}$ be Banach spaces with uniformly convex duals, then there is an $\varepsilon_o > 0$ such that for all $\varepsilon < \varepsilon_o$ the following implication holds:

if T is a linear isomorphism from $X \check{\otimes} \tilde{X}$ onto itself with $\|T\|\cdot\|T^{-1}\| \leq 1+\varepsilon$ and if there exist $x_o^* \in E(X^*)$ and $\tilde{x}_o^* \in E(\tilde{X}^*)$ such that

$$T^*(x_o^* \otimes \tilde{x}^*) = x_o^* \otimes \tilde{x}^* \quad \text{for all} \quad \tilde{x}^* \in \tilde{X}^*$$

and

$$T^*(x^* \otimes \tilde{x}_o^*) = x^* \otimes \tilde{x}_o^* \quad \text{for all} \quad x^* \in X^*$$

then $\|T - Id\| \leq 6\gamma(\varepsilon)$.

<u>Proof of the lemma</u>. Let $x_1^* \in E(X^*)$, $\tilde{x}_1^* \in E(\tilde{X}^*)$. It follows from the assumption and our previous considerations that there are isomorphisms $\Phi \in L(X)$ and $\Psi \in L(X)$ such that

$$\|T^*(x_1^* \otimes \tilde{x}^*) - x_1^* \otimes \Phi^*(\tilde{x}^*)\| \leq 2\gamma(\varepsilon) \quad \text{for all} \quad \tilde{x}^* \in E(\tilde{X}^*)$$

and

$$\|T^*(x^* \otimes \tilde{x}_1^*) - \Psi^*(x^*) \otimes \tilde{x}_1^*\| \leq 2\gamma(\varepsilon) \quad \text{for all} \quad x^* \in E(X^*).$$

Substituting $\tilde{x}^* = \tilde{x}_1^*$ and $x^* = x_1^*$ we get

$$\|T^*(x_1^* \otimes \tilde{x}_1^*) - x_1^* \otimes \Phi^*(\tilde{x}_1^*)\| \leq 2\gamma(\varepsilon)$$

and

$$\|T^*(x_1^* \otimes \tilde{x}_1^*) - \Psi^*(x_1^*) \otimes \tilde{x}_1^*\| \leq 2\gamma(\varepsilon).$$

Hence $\|\Phi^*(\tilde{x}_1^*) - \tilde{x}_1^*\| \leq 4\gamma(\varepsilon)$ and $\|x_1^* - \Psi^*(x_1^*)\| \leq 4\gamma(\varepsilon)$, so

$\|T^*(x_1^* \otimes \tilde{x}_1^*) - x_1^* \otimes \tilde{x}_1^*\| \leq 6\gamma(\varepsilon)$ ending the proof of the lemma.

●

The results of this section are based on the author's papers [6,9].

14. Introduction.

In the previous chapters we have proven, very roughly speaking, ıat in various classes of Banach algebras the metric and the alge-·aic perturbations coincide. In this chapter we use these results, ›stly Theorem 3.1, to investigate how the algebraic properties of ınction algebras change under small perturbations. We are specially ıterested in the properties which are stable, i.e. which are inva-.ant under small perturbations.

Let $\mathcal{A}$ be a class of all function algebras modulo the equivalence ·lation: A is equivalent to B if the Banach-Mazur distance between and B is equal to one. Example 3.10 shows that this relation is ›t trivial, i.e. there are function algebras A and B which are ˌuivalent but not equal (in the category of Banach algebras).

According to Rochberg's papers we define a function D on $\mathcal{A}\times\mathcal{A}$ ıich is, from the point of view of perturbations a natural measure of .osedness of two uniform algebras A and B:

$$D(A,B) = \inf\{\varepsilon:\ B \text{ is algebraically isomorphic to an algebraic } \varepsilon\text{-perturbation of } A\}.$$

: we regard A and B as a Banach spaces a natural measure of closed-·ss is logarithm of the Banach-Mazur distance:

$$d(A,B) = \inf\{\log \|T\|\,\|T^{-1}\| : T \in L(A,B)\}.$$

As an immediate consequence of Theorem 3.1 we get that these two :e equivalent.

14.1. Proposition. Let $B, A_1, A_2, \ldots$ be function algebras then

$$\lim_{n\to\infty} D(B,A_n) = 0 \quad \text{if and only if} \quad \lim_{n\to\infty} d(B,A_n) = 0.^{(1)}$$

(1) Inspection of the proof of Theorem 3.1 actually shows that there are

It is known that $d(\cdot,\cdot)$ is an extended valued metric on $\mathcal{A}$. In this section we study the metric space $(\mathcal{A},d)$ and its various subsets.

14.2. Definition. Let (P) be a property which is defined for every function algebra. We call (P) stable if for any function algebra A which has (P) there is a positive number ε such that for any ε-deformation B of A the algebra B has (P).

Notice that a property (P) is stable if and only if the subset of $\mathcal{A}$ consisting of all function algebras which have (P) is an open subset of $(\mathcal{A},d)$. In Chapter I we had some examples of stable properties.

1. For any Hausdorff space S the property "ChA is homeomorphic to S" is stable (Theorem 3.1(v)).
2. The property "$\partial A = \mathrm{Ch}A$" is stable (Corollary 3.12).
3. For any compact Hausdorff space S the algebra C(S) is an isolated point of $\mathcal{A}$, i.e. the property "A = C(S)" is stable (Corollary 3.14).

Later on we will prove that also the following properties are stable.

4. "A is a Dirichlet algebra".
5. "A is antisymmetric".
6. "$\mathfrak{M}(A)$ has exactly n components", for $n \in \mathbb{N} \cup \{\infty\}$.

The subsets of $\mathcal{A}$ defined by properties 1 - 6 are even closed and open in $\mathcal{A}$.

In the sequel the elements of $\mathcal{A}$ will be denoted by the corresponding function algebras (when there is no danger of ambiguity).

§ 15. $\mathcal{A}$ space.

15.1. THEOREM (Rochberg). $\mathcal{A}$ is a complete metric space.

---

positive constants ε_0 and c such that for any function algebras A and B we have

$$\frac{1}{c}\, d(A,B) \le D(A,B) \le c \cdot d(A,B)$$

whenever $\min (d(A,B), D(A,B)) \le \varepsilon_0$.

Proof. Let (A_n) be a Cauchy sequence in $\mathcal{A}$. To show that (A_n) is convergent it is sufficient to show it has a convergent subsequence. So by Corollary 3.7a we can assume there is a sequence of linear isomorphisms $T_n: A_{n-1} \to A_n$ such that $T_n(\mathbf{1}) = \mathbf{1}$ and

$$\|T_n\| \|T_n^{-1}\| \le 1 + \frac{1}{2^n} \qquad \text{for any} \quad n \in \mathbb{N}.$$

Put

$$S_n = T_{n-1} \circ T_{n-2} \circ \ldots \circ T_1 : A_1 \to A_n.$$

We define a second norm $|||\cdot|||$ on A_1 by

$$|||f||| = \lim_{n\to\infty} \|S_n(f)\| \qquad \text{for} \quad f \in A_1.$$

A direct verification shows that $|||\cdot|||$ is a well-defined, complete norm on A_1, equivalent to the original one and that

$$d((A_1, |||\cdot|||), A_n) \xrightarrow[n\to\infty]{} 0.$$

By Corollary 3.7b for any $f, g \in A_1$ the sequence $(S_n^{-1}(S_n f \cdot S_n g))$ is a Cauchy sequence and we can define a second multiplication $\times$ on A_1 by

$$f \times g = \lim_{n\to\infty} S_n^{-1}(S_n f \cdot S_n g) \qquad \text{for} \quad f,g \in A_1.$$

It is easy to check that $(A_1, |||\cdot|||, \times)$ is a uniform algebra and this ends the proof.

●

Before stating the further theorems let us recall some notation. For a uniform algebra A let

- $A^{-1} = \{f \in A: \ f^{-1} \in A\}$,
- $e^A = \{f \in A: \ \exists g \in A, \ f = \exp(g)\}$,
- $\mathrm{Re}A = \{\mathrm{Re}f: \ f \in A\}$,
- $C_R(\partial A)$ be the space of continuous real-valued functions on ∂A with the uniform norm,
- $\overline{\mathrm{Re}A}$ be the closure of $\mathrm{Re}A$ in $C_R(\partial A)$,
- $X(A) = C_R(\partial A)/\overline{\mathrm{Re}A}$.

15.2. THEOREM (Rochberg). For any $n \in \mathbb{N} \cup \{\infty\}$ the subset of $\mathcal{A}$ consisting of all function algebras such that $\dim X(A) = n$ is closed and open.

For the proof we need the following propositions.

15.3. Proposition (Rochberg). For any positive constant K there is an $\varepsilon_o > 0$ such that for any function algebras A,B and any onto isomorphism $T: A \to B$ with $T(\mathbb{1}) = \mathbb{1}$ and $\|T\|\cdot\|T^{-1}\| \leq 1 + \varepsilon_o$ we have

a) if $f \in A^{-1}$ and $\|f\|\cdot\|f^{-1}\| \leq K$ then $Tf \in B^{-1}$;

b) if $f \in e^A$ and $\|f\|\cdot\|f^{-1}\| \leq K$ then $Tf \in e^B$.

Proof. By Corollary 3.7b we have

$$\|T(f) \cdot T(f^{-1}) - \mathbb{1}\| \leq O(\varepsilon)\,\|f\|\cdot\|f^{-1}\| \leq O(\varepsilon) \cdot K$$

and hence $T(f) \cdot T(f^{-1}) \in B^{-1}$ if ε is sufficiently small. To prove b) we recall that e^B is the connected component of B^{-1} which containes $\mathbb{1}$. Assume $f = e^h \in e^A$ and $\|f\|\cdot\|f^{-1}\| \leq K$. For $0 \leq t \leq 1$ we have

$$\|e^{th}\|\cdot\|e^{-th}\| \leq e^{2K} \quad .$$

Hence by a) we have $T(e^{th}) \in B^{-1}$ and $[0,1] \ni t \mapsto T(e^{th})$ is a continuous function so $T(e^h) \in e^B$.

●

The following is a well-known fact.

15.4. Proposition. Let X be a Banach space and let X_1, X_2 be closed subspaces of X. Assume there is a $c < 1$ such that for any $f_1 \in X_1$ and $f_2 \in X_2$ with $\|f_1\| = 1 = \|f_2\|$ there are $g_1 \in X_2$ and

$g_2 \in X_1$ with $\|g_1\| = 1 = \|g_2\|$ such that

$$\|f_1 - g_1\| \leq c \quad \text{and} \quad \|f_2 - g_2\| \leq c,$$

then $\dim X/X_1 = \dim X/X_2$.

To end the proof of Theorem 15.2 it is sufficient, by the above Proposition and Theorem 3.1, to show that there is an $\varepsilon_o > 0$ such that:

If A,B are closed subalgebras of $C(S)$, with S a Hausdorff space, and if there is a linear onto isomorphism $T: A \to B$ such that

$$T(\mathbb{1}) = \mathbb{1} \quad \text{and} \quad \|T(f) - f\| \leq \varepsilon_o \|f\| \quad \text{for} \quad f \in A, \tag{62}$$

then for any $\mathrm{Re} f \in \mathrm{Re} A$ with $\|\mathrm{Re} f\| = 1$ there is a $\mathrm{Re} g \in \mathrm{Re} B$ such that

$$\|\mathrm{Re} f - \mathrm{Re} g\| \leq \frac{1}{3}.$$

To this end put $K = e^2$ and let ε_o be as in Proposition 15.3. We assume additionally that $\varepsilon_o \leq 0.05$. Fix $f \in A$ with $\|\mathrm{Re} f\| = 1$. We have

$$\|e^{\pm f}\| = \sup\{e^{\pm \mathrm{Re}\, f(s)} : s \in S\} \leq e$$

so

$$\|e^f\| \cdot \|e^{-f}\| \leq e^2.$$

By Proposition 15.3 there is a $g \in B$ such that $T(e^f) = e^g$. From (62) we get

$$\varepsilon_o e \geq \|T(e^f) - e^f\| = \|e^g - e^f\| = \|e^f(\mathbb{1} - e^{g-f})\| \geq \frac{1}{e}\|\mathbb{1} - e^{g-f}\|.$$

Hence

$$\|e^{g-f}\| \leq 1 + \varepsilon_o e^2 \quad \text{and} \quad \|e^{f-g}\| \leq \frac{1}{1-\varepsilon_o e^2}$$

and this gives

$$\|\mathrm{Re} f - \mathrm{Re} g\| \leq \max\{\ln(1+\varepsilon_o e^2), |\ln(1-\varepsilon_o e^2)|\} < \frac{1}{3}.$$

∎

Recall that a uniform algebra A is called a Dirichlet algebra if $C_R(\partial A) = \overline{\mathrm{Re}A}$.

15.5. Corollary. There is an $\varepsilon_o > 0$ such that any ε_o-deformation of a Dirichlet algebra is a Dirichlet algebra.

§ 16. The decomposition of a deformation into antisymmetric algebras.

Let us recall the well-known Shilov-Bishop theorem.

Theorem (Shilov-Bishop). Let A be a function algebra. Then A is a sum of the family of disjoint, maximal sets of antisymmetry $\{S_i\}_{i\in I}$ of A and

$$f \in A \quad \text{if and only if} \quad f \in C(\partial A) \quad \text{and} \quad f|_{S_i} \in A|_{S_i} \quad \text{for } i \in I.$$

•

In this section we prove that for any function algebra A and its any small perturbation B we have: for any maximal set of antisymmetry S of A there is a maximal set of antisymmetry S' of B such that $B|_{S'}$ is a small perturbation of $A|_S$. As a corollary we get that the property "A is antisymmetric" is stable. We also get, via the Shilov-Bishop theorem, a method which allows to restrict investigation of small perturbations of function algebras to investigation of perturbations of the antisymmetric algebras.

To prove our "decomposition theorem" we need the following generalization of the Bishop's so-called "$\frac{3}{4} - \frac{1}{4}$" criterion.

For this purpose let us recall the definition of a (weak) peak set for a function algebra.

16.1. Definition. Let A be a function algebra defined on a compact Hausdorff space S. A closed subset S_o of S is called a (weak) peak set for A if (for any open neighbourhood U of S) there exists an f in A such that $1 = \|f\| = f(s)$ for $s \in S_o$ and $|f(s)| < 1$ for $s \in S \setminus S_o$ (for $s \in S \setminus U$).

•

Notice that if S is metric then any weak peak set for A is a peak set.

16.2. THEOREM. Let A be a function algebra on a compact Hausdorff space S and let S_o be a closed subset of S. Then S_o is a weak peak set for A if and only if there are constants $K \geq 1$, $c_1 \geq 0$, $c_2 \geq 0$ with $c_1 + c_2 < 1$ such that for any open neighbourhood U of S_o there is an f in A such that $\|f\| \leq K$, $|f(s) - 1| \leq c_2$ for $s \in S_o$ and $|f(s)| \leq c_1$ for $s \in S \setminus U$.

Note that the definition of a weak peak set is evidently equivalent to the following one.

Let A be a function algebra on S and let S_o be a closed subset of S, then S_o is a weak peak set for A if and only if for any $\varepsilon > 0$ and any open neighbourhood U of S_o there exists an f in A such that $1 = \|f\| = f(s)$ for $s \in S_o$ and $|f(s)| < \varepsilon$ for $s \in S \setminus U$.

Hence, the "only if" part of our theorem is trivial, i.e. if $S_o \subset S$ is a weak peak set for A then as c_1, c_2 we can take any positive numbers and put $K = 1$.

We devide the proof of our theorem into four steps. Assume that the assumptions of Theorem are fulfilled.

Step 1. For any $\varepsilon > 0$ there is a positive constant $K(\varepsilon)$ such that for any open neighbourhood U of S_o there is an f in A such that $|f(s)| \leq \varepsilon$ for $s \in S \setminus U$; $|f(s)| \leq 1$ and $|f(s) - 1| \leq \varepsilon$ for $s \in S_o$ and $\|f\| \leq K(\varepsilon)$.

Proof. Let $f_1 \in A$ be such that $|f_1(s)| \leq c_1$ for $s \in S \setminus U$, $|f_1(s) - 1| \leq c_2$ for $s \in S_o$ and $\|f_1\| \leq K$. Since the discs $D_1 = \{z \in \mathbb{C}: |z| \leq c_1\}$ and $D_2 = \{z \in \mathbb{C}: |z - 1| \leq c_2\}$ are disjoint then, by Runge Theorem, there is a polynomial p such that $p(z) \leq \varepsilon$ for $z \in D_1$ and $|p(z) - (1 - \frac{\varepsilon}{2})| \leq \frac{\varepsilon}{2}$ for $z \in D_2$. Put $K(\varepsilon) = \sup\{|p(z)|: |z| \leq K\}$ and $f = p \circ f_1 \in A$.

•

Step 2. Assume that there are constants K_1 and $\varepsilon < 1$ such that for any open neighbourhood U of S_o there is an f in A such that

$$|f(s)| \leq \varepsilon \quad \text{for} \quad s \in S \setminus U; \quad \|f\| \leq K_1;$$

$$|f(s)| \leq 1 \quad \text{and} \quad |f(s) - 1| \leq \varepsilon \quad \text{for} \quad s \in S_o .$$

then for any open neighbourhood U of S_o there is a g in A such that

$$|g(s)| \leq \varepsilon \quad \text{for} \quad s \in S \setminus U; \quad \|g\| \leq 1;$$

$$|g(s) - 1| \leq \varepsilon \quad \text{for} \quad s \in S_o.$$

•

Proof. Fix any $x < 1$ with

$$a = (K_1 - 1) - x(K_1 - \varepsilon) < 0$$

and a decreasing sequence of positive numbers ε_n such that

$$\varepsilon_n(1 - x^n) + x^n a < 0 \qquad \text{for} \quad n \geq 1.$$

We define by induction a sequence of functions $(h_n)_{n=1}^{\infty}$ from A. Let $h_1 \in A$ be any function from A such that

$$|h_1(s)| \leq \varepsilon \quad \text{for} \quad s \in S \setminus U; \quad \|h_1\| \leq K_1;$$

$$|h_1(s)| \leq 1 \quad \text{and} \quad |h_1(s) - 1| \leq \varepsilon \quad \text{for} \quad s \in S_o.$$

Assume we have defined $h_1, \ldots, h_n$ then put

$$W_n = \{s: \max_{1 \leq j \leq n} |h_j(s)| \geq 1 + \varepsilon_n\}.$$

The set W_n is a closed subset of $S \setminus S_o$ so there is an $h_{n+1} \in A$ such that

$$|h_{n+1}(s)| \leq \varepsilon \quad \text{for} \quad s \in (S \setminus U) \cup W_n; \quad \|h_{n+1}\| \leq K_1;$$

$$|h_{n+1}(s)| \leq 1 \quad \text{and} \quad |h_{n+1}(s) - 1| \leq \varepsilon \quad \text{for} \quad s \in S_o.$$

Let

$$g = (1 - x) \sum_{j=1}^{\infty} x^{j-1} \cdot h_j .$$

We have evidently $|g(s)| \leq 1$ and $|g(s) - 1| \leq \varepsilon$ for $s \in S_0$, $|g(s)| \leq \varepsilon$ for $s \in S \setminus U$ and $|g(s)| \leq 1$ for $s \in S \setminus \bigcup_{n=1}^{\infty} W_n$. It remains to show that if $s \in \bigcup_{n=1}^{\infty} W_n$ then $|g(s)| \leq 1$.

The sequence $(W_n)_{n=1}^{\infty}$ is an increasing sequence of compact sets so if $s \in \bigcup_{n=1}^{\infty} W_n$ then there is a positive integer m such that $s \in W_{m+1}$ but $s \notin W_m$ for $m \geq 0$ (we put $W_0 = \emptyset$). We have

$$|h_j(s)| \leq 1 + \varepsilon_m \quad \text{for} \quad j \leq m,$$

$$|h_{m+1}(s)| \leq K_1,$$

$$|h_j(s)| \leq \varepsilon \quad \text{for} \quad j \geq m + 2$$

hence

$$|g(s)| \leq (1 - x)\cdot[(1 + \varepsilon_m)\cdot\sum_{j=1}^{m} x^{j-1} + K_1 x^m + \varepsilon \sum_{j=m+2}^{\infty} x^{j-1}] =$$

$$= 1 + \varepsilon_m(1 - x^m) + x^m(K_1 - 1 - x(K_1 - \varepsilon)) < 1.$$

•

<u>Step 3</u>. Let M be a closed subspace of $C(S)$, let π be a canonical map from M onto $M|_{S_0} = \{f|_{S_0} : s \in M\} \subset C(S_0)$, $\pi(f) = f|_{S_0}$ and let $\tilde{\pi}: {}^{M}/_{\ker \pi} \to M|_{S_0} : \tilde{\pi}(f + \ker \pi) = f|_{S_0}$. Then if $\tilde{\pi}$ is not an isometry then there are measures μ on S_0 and ν on $S \setminus S_0$ such that $\mu - \nu \perp M$ and μ represents a non zero functional on M.

<u>Proof</u>. Assume $\tilde{\pi}$ is not an isometry, there is a functional F_0 on $M|_{S_0}$ represented by a measure μ_0 on S_0 such that $\|F_0\| = \text{var}(\mu_0) = 1$ but $\|\pi^*(F_0)\| = t < 1$. Let ν_0 be any measure on S which represents the functional $\pi^*(F_0)$ with $\text{var}(\nu_0) = t$. Since $\text{var}(\nu_0|_{S_0}) \leq t < 1$ then the measure $\mu_0 - \nu_0|_{S_0}$ is not orthogonal to M and we can put $\mu = \mu_0 - \nu_0|_{S_0}$, $\nu = \nu_0|_{S \setminus S_0}$.

•

To end the proof of Theorem, now fix an open neighbourhood U of S_0 and let $q \in C_R(S)$ be such that

$$q|_{S_0} \equiv 1, \quad q|_{S \setminus U} \equiv 3, \quad 1 \leq q \leq 3$$

and put

$$M = \{q \cdot g \in C(S): \quad g \in A\}.$$

We shall prove that $\tilde{\pi}: {}^{M}/_{\ker \pi} \to M|_{S_0}$ is an isometry. Assume to the contrary and let μ, ν be as in Step 3. We can assume that the norm of μ on $M|_{S_0}$ is equal to 2 and let $f \in M$ be such that $\int f \, d\mu = 1$ and $\|f\|_{S_0} \leq 1$. Fix $0 < \delta < \frac{1}{2}(\|f\|(1 + \mathrm{var}(\nu)))^{-1}$. The regularity of the measure ν provides there is an open neighbourhood V of S_0 such that $|\nu|(V) < \delta$ and, by Steps 1 and 2, there is an f_0 in A such that $\|f_0\|_{S \setminus V} \leq \delta$, $\|f_0 - 1\|_{S_0} \leq \delta$ and $\|f_0\| \leq 1$. Since $\nu - \mu \perp M$ and $ff_0 \in M$ we have

$$\frac{1}{2} < 1 - 2\delta \leq \left|\int_{S_0} f d\mu\right| - \left|\int_{S_0} f(1 - f_0) d\mu\right| \leq$$

$$\leq \left|\int_{S_0} ff_0 d\mu\right| = \left|\int_{S \setminus S_0} ff_0 d\nu\right| \leq \left|\int_V ff_0 d\nu\right| + \left|\int_{S \setminus V} ff_0 d\nu\right| \leq$$

$$\leq \delta\|f\| + \mathrm{var}(\nu)\delta\|f\| < \frac{1}{2}.$$

So we proved that $\tilde{\pi}$ is an isometry and the same time we established that:

for any open neighbourhood U of S_0 there is an f_U in A such that $\|f_U\| \leq 2$, $f_U|_{S_o} \equiv 1\!\!1$ and $|f_U(s)| \leq \frac{1}{2}$ for $s \in S \setminus U$.

Now Theorem follows from the Bishop's criterion applied to the algebra $\tilde{A} = \{f \in A: f|_{S_0} = \text{const}\}$.

Before stating further theorems observe that for any weak peak set S_0 for a function algebra A the set S_0 is uniquely determined by $S_0 \cap \mathrm{Ch}A$:

$$S_o = \mathcal{cl}(S_0 \cap \mathrm{Ch}A) \tag{63}$$

where by $\mathcal{cl}(\cdot)$ we denote the closure in the hull-kernel topology defined by

$$\mathcal{cl}(F) = \{s \in \partial A: \quad \forall f \in A \quad [f|_F \equiv 0 \Rightarrow f(s) = 0]\}.$$

16.3. THEOREM. There is an $\varepsilon_0 > 0$ such that for any function algebras A and B and any linear onto isomorphism $T: A \to B$ with $T(\mathbf{1}) = \mathbf{1}$ and $\|T\| \cdot \|T^{-1}\| \leq 1 + \varepsilon_0$ there is a one to one correspondence between the weak peak sets for A and the weak peak sets for B:

$Y_0 = \bar{Y}_0 \subset \partial B$ is a weak peak set for B if and only if

$cl(\varphi(Y_0 \cap ChB)) \subset \partial A$ is a weak peak set for A;

where φ is a homeomorphism between ChB and ChA given for T, by Theorem 3.1 (φ).

Proof. Let $Y_0 = \bar{Y}_0 \subset \partial B$ be a weak peak set for B. Put

$$X_0 = \{x_0 \in \partial A: \text{ there is a net } \{y_\alpha\}_{\alpha \in \Gamma} \text{ of elements of ChB tending to some element of } Y_0 \text{ and such that } \varphi(y_\alpha) \to x_0\}.$$

Fix $\varepsilon > 0$. Let $(U_\alpha)_{\alpha \in \Gamma}$ be a family of open neighbourhoods of $Y_0 \subset \partial B$ and let $(g_\alpha)_{\alpha \in \Gamma}$ be the corresponding family of functions from B such that

$$g_\alpha\big|_{Y_0} \equiv 1 = \|g_\alpha\| \text{ and } |g_\alpha(y)| \leq \varepsilon \quad \text{for} \quad y \in \partial B \setminus U_\alpha, \alpha \in \Gamma.$$

Put $f_\alpha = T^{-1} g_\alpha$. Hence, by Theorem 3.1, we have

$$\left.\begin{array}{l} \|f_\alpha\| \leq 1 + \varepsilon_0, \quad |f_\alpha(x) - 1| \leq \varepsilon_6 \quad \text{for} \quad x \in X_0 \\ |f_\alpha(x)| \leq \varepsilon + \varepsilon_6 \quad \text{for} \quad x \in K_\alpha, \end{array}\right\} \tag{64}$$

and

where $K_\alpha = \overline{ChA \setminus \varphi(U_\alpha \cap ChB)}$ and ε_6 is as in Theorem 3.1. The sets X_0 and K_α, $\alpha \in \Gamma$ are compact and it is easy to check that $\bigcup_\alpha K_\alpha \cup X_\alpha = \partial A$. In view of (64), by Theorem 16.2 it can be proven that X_0 is a weak peak set for A (if $2\varepsilon_6 < 1$), so (Larsen [1]) X_0 is the closure in the hull-kernel topology of $X_0 \cap ChA = \varphi(Y_0 \cap ChB)$, which by symmetry ends the proof.

●

16.4. THEOREM. There is an $\varepsilon_o > 0$ such that for any function algebras A and B, if the Banach-Mazur distance between A and B is not greater than $1 + \varepsilon_o$ and A is antisymmetric then B is also an antisymmetric algebra.

As usual, we call a function algebra A antisymmetric if there is no non-constant real valued functions in $A \subset C(\partial A)$.

Proof. By Corollary 3.7a) we can assume, without loss of generality, that there is an onto isomorphism $T: A \to B$ with $T\mathbb{1} = \mathbb{1}$ and $\|T\| \cdot \|T^{-1}\| \leq 1 + \varepsilon_o$. Let φ be a homeomorphism from ChB onto ChA given by Theorem 3.1.

Assume B is not antisymmetric and let g_o be a non-constant real-valued function from B. We define an equivalence relation on ∂B:

$$y_1 \sim y_2 \quad \text{if and only if} \quad g_o(y_1) = g_o(y_2).$$

Let π be a natural projection from ∂B onto a compact space $\partial B/_{\sim} = K$. By the Stone-Weierstrass theorem for any $g \in C(K)$ we have $g \circ \pi \in B$, so for any compact subset K_o of K the set $\pi^{-1}(K_o)$ is a weak peak set for B. Hence, by Theorem 16.3, for any $k_1 \neq k_2 \in K$ sets $cl\{\varphi(\pi^{-1}(k_1))\}$ and $cl\{\varphi(\pi^{-1}(k_2))\}$ are weak peak sets, hence $S = cl\{\varphi(\pi^{-1}(k_1))\} \cap cl\{\varphi(\pi^{-1}(k_2))\}$ is also a weak peak set. But $S \cap \mathrm{ChA} = \emptyset$ so $S = \emptyset$. Hence φ and π induce a continuous map $\tilde{\pi}$ from ∂A onto K such that

$$\pi(y) = \tilde{\pi}(\varphi(y)) \qquad \text{for all} \qquad y \in \mathrm{ChB}$$

and having the property that $\tilde{\pi}^{-1}(K_o)$ is a weak peak set for any closed subset K_o of K.

Let $\tilde{g}_o \in C(K)$ be such that $g_o = \tilde{g}_o \circ \pi$ and put $f_o = \tilde{g}_o \circ \tilde{\pi}$. Since A is antisymmetric, $f_o \notin A$ so there exists a measure μ on ∂A orthogonal to A such that $\mu(f_o) \neq 0$. Since $\mu(f_o) \neq 0$ and f_o is of the form $\tilde{g}_o \circ \tilde{\pi}$ there is a $K_o = \overline{K}_o \subset K$ for which $\mu(\tilde{\pi}^{-1}(K_o)) \neq 0$; we can assume $\mu(\tilde{\pi}^{-1}(K_o)) = 1$. Let $U \subset \partial A$ be an open neighbourhood of $\tilde{\pi}^{-1}(K_o)$ with $|\mu|(U \setminus \tilde{\pi}^{-1}(K_o)) \leq \frac{1}{4}$. Since $\tilde{\pi}^{-1}(K_o)$ is a weak peak set there exists an $f \in A$ such that

$$\|f\| = 1 \equiv f\Big|_{\pi^{-1}(K_o)} \quad \text{and} \quad |f(x)| \le \frac{1}{4\mathrm{var}(\mu)} \quad \text{for} \quad x \in \partial A \setminus U.$$

We have

$$0 = |\int f d\mu| \ge |\int_{\pi^{-1}(K_o)} f d\mu| - |\int_{U\setminus \pi^{-1}(K_o)} f d\mu| - |\int_{\partial A\setminus U} f d\mu| \ge$$

$$\ge 1 - \frac{1}{4} - \frac{1}{2\mathrm{var}(\mu)} |\mu|(\partial A\setminus U) \ge \frac{1}{4}$$

and the above contradiction ends the proof.

•

16.5. Remark. Notice that in fact we do not use in the proof of the above Theorem the assumption that the Banach-Mazur distance between A and B is small but we use only the following, weaker assumption:

There is a homeomorphism φ from ChB onto ChA such that

$S_o \subset \partial B$ is a weak peak set for B if and only if

$cl(\varphi(S_o \cap \mathrm{Ch}B)) \subset \partial A$ is a weak peak set for A.

For the Theorem which follows let us recall a function algebra A is termed stable if there is an $\varepsilon > 0$ such that for any ε-perturbation $\times$ of A, the algebras A and $(A,\times)$ are isomorphic.

16.6. THEOREM. Assume A and B are stable function algebras then $A \oplus B$ - the direct sum of A and B is also stable.

Proof. Let C be a function algebra and let $T: A \oplus B \to C$ be an onto isomorphism with $T\mathbb{1} = \mathbb{1}$ and $\|T\|\cdot\|T^{-1}\| \le 1 + \varepsilon$; where $\varepsilon > 0$ is smaller than the constants ε_o in Theorems 3.1 and 16.3. Let $\varphi: \mathrm{Ch}C \to \mathrm{Ch}(A \oplus B) = \mathrm{Ch}A \cup \mathrm{Ch}B$ be as in Theorem 3.1. By Theorem 16.3 the sets $Y_1 = \overline{\varphi^{-1}(\mathrm{Ch}A)}$ and $Y_2 = \overline{\varphi^{-1}(\mathrm{Ch}B)}$ are weak peak sets so there is a $g \in C$ such that $g|_{Y_1} \equiv 1$ and $g|_{Y_2} \equiv 0$ hence C is a direct sum of the algebras $C_1 = g \cdot C = \{g \cdot h \in C: h \in C\}$ and $C_2 = (1-g) \cdot C$.

Put

$T_1: A \to C_1: \; T_1(f) = T(f) \cdot g \qquad \text{for} \quad f \in A,$

$$T_2 : B \to C_2 : \quad T_2(f) = T(f) \cdot (1 - g) \qquad \text{for} \quad f \in B.$$

By Theorem 3.1 (φ) it follows that T_1 and T_2 are onto isomorphisms with

$$\|T_i\| \, \|T_i^{-1}\| \leq 1 + \varepsilon'(\varepsilon) \qquad \text{for} \quad i = 1,2.$$

Hence, if ε is sufficiently small, the stability of A and B implies that A and C_1 as well as B and C_2 are isomorphic, so $A \oplus B$ and $C = C_1 \oplus C_2$ are isomorphic (in the category of algebras).

•

Now we are ready to summarize the results of this section in Decomposition Theorem.

16.7. THEOREM. There is a positive constant ε_o such that for any function algebras A and B and any linear isometry T from A onto B with $\|T\| \, \|T^{-1}\| \leq 1 + \varepsilon$ and $T(1\!\!1) = 1\!\!1$, where $\varepsilon \leq \varepsilon_o$ we have

a) there is a homeomorphism φ from ChB onto ChA such that

$$\|T(f)(s) - f \circ \varphi(s)\| \leq \varepsilon'(\varepsilon) \|f\| \qquad \text{for} \quad f \in A, \quad s \in \text{ChB};$$

b) a closed subset S_o of ∂B is a (weak) peak set for B if and only if $cl(\varphi(S_o \cap \text{ChB}))$ is a (weak) peak set for A;

c) a closed subset S_o of ∂B is a maximal set of antisymmetry of B if and only if $cl(\varphi(S_o \cap \text{ChB}))$ is a maximal set of antisymmetry of A.

d) if S_o is a metrizable maximal set of antisymmetry of B then

$$d(B|_{S_o}, A|_{cl(\varphi(S_o \cap \text{ChB}))}) \leq \varepsilon'(\varepsilon),$$

where $\lim_{\varepsilon \to 0} \varepsilon'(\varepsilon) = 0$.

Proof. Points a) and b) follow from Theorem 3.1 (φ) and Theorem 16.3, respectively.

To prove c) we recall that any maximal set of antisymmetry of a function algebra A is a weak peak set for A and that if S_o is a weak peak set for A, and K is a weak peak set for $A|_{S_o}$ then K is a weak peak set for A. To prove c), by symmetry, it is sufficient to show that if S_o is a maximal set of antisymmetry of B then $cl(\varphi(S_o \cap ChB))$ is a set of antisymmetry of A. To this end put

$$A' = A|_{cl(\varphi(S_o \cap ChB))} \quad , \quad B' = B|_{S_o}.$$

It is easy to verify that

$$\partial A' \subset cl(\varphi(S_o \cap ChB)), \qquad \partial B' \subset S_o,$$
$$ChA' = \varphi(S_o \cap ChB), \qquad ChB' = S_o \cap ChB.$$

Hence, by b), $\tilde{\varphi} = \varphi|_{S_o \cap ChB}$ is a homeomorphism from ChB' onto ChA' such that

$K \subset S_o$ is a weak peak set for B' if and only if $cl(\varphi(K \cap ChB'))$ is a weak peak set for A'.

Point c) follows now from Remark 16.4 applied to the algebras A' and B'.

To prove d) we need the following theorem of A.Pełczyński [1].

Theorem (Pełczyński). Let A be a function algebra on a compact Hausdorff space S and let S_o be a metrizable weak peak set for A then there is a linear extension of norm one from $A|_{S_o}$ into A, this means there is a norm one linear map $\Phi: A|_{S_o} \to A$ such that

$$\Phi(f)|_{S_o} \equiv f \qquad \text{for any} \qquad f \in A|_{S_o}.$$

To end the proof of d) let S_o be a metrizable maximal set of antisymmetry of B and let $\Phi: B|_{S_o} \to B$ be a norm one linear extension. We define

$$T_1: B|_{S_o} \to A|_{cl(\varphi(S_o \cap ChB))}$$

by

$$T_1(g) = T^{-1} \circ \Phi(g)|_{cl(\varphi(S_o \cap ChB))}.$$

From a) we have

$$(1 - 2\varepsilon')\|g\| \le \|T(g)\| \le (1 + 2\varepsilon')\|g\| \qquad \text{for} \quad g \in B|_{S_o}. \tag{65}$$

Fix $f \in A|_{cl(\varphi(S_o \cap ChB))}$ with $\|f\| = 1$ and let $\tilde{f} \in A$ be such that $\|\tilde{f}\| = 1$ and $\tilde{f}|_{cl(\varphi(S_o \cap ChB))} \equiv f$. From a) we get

$$|T_1(T(f))(s) - f(s)| \leq 3\varepsilon' \quad \text{for any} \quad s \in cl(\varphi(S_o \cap ChB)).$$

Since f is an arbitrary element of $A|_{cl(\varphi(S_o \cap ChB))}$ with norm one this proves that $T_1(B|_{S_o})$ is a dense subspace of $A|_{cl(\varphi(S_o \cap ChB))}$ (if $3\varepsilon' < 1$).

By (65) T_1 is a closed map so we get that T_1 is an onto map and

$$\|T_1\| \, \|T_1^{-1}\| \leq 1 + 5\varepsilon'.$$

•

§ 17. A non-stable Dirichlet algebra.

In this section we investigate the following problem arising naturally from Decomposition Theorem.

17.1. Problem. Let A be a function algebra and let

$$\mathcal{F} = \{A|_S : \ S \text{ is a maximal set of antisymmetry of } A\}$$

be a uniformly stable family of function algebras i.e. assume that there is an $\varepsilon > 0$ such that for any $A' \in \mathcal{F}$ and any ε-perturbation $\times$ of A', the algebras A' and $(A', \times)$ are isomorphic. Is then A stable?

•

The answer is negative. The algebra which we construct below is Dirichlet, so the example gives also a negative answer to the question whether all Dirichlet algebras are stable.

The following two properties of the disc algebra $A(D)$ are crucial in our example.

1. $A(D)$ is stable.
2. $A(D)$ is not strongly stable.

The stability of the disc algebra will be discussed in the next section in the wider context of algebras of analytic functions on

Riemann surfaces. Now we prove 2.

17.2. Definition. A function algebra A is called strongly stable if there is an $\varepsilon_o > 0$ such that for any $\varepsilon < \varepsilon_o$, any function algebra B, and any linear isomorphism T from A onto B with $\|T\|\,\|T^{-1}\| \leq 1 + \varepsilon$ and $T(\mathbf{1}) = \mathbf{1}$ there is an onto algebra isomorphism $\tilde{T}: A \to B$ such that $\|\tilde{T} - T\| \leq \varepsilon'(\varepsilon)$, and $\lim_{\varepsilon \to 0} \varepsilon'(\varepsilon) = 0$.

•

17.3. Example. By Theorem 3.1 we get, as in the proof of Corollary 3.14, that the algebra $C(S)$ is strongly stable for any compact Hausdorff space S.

•

17.4. THEOREM (Johnson). The disc algebra $A(D)$ is not strongly stable.

Proof (Rochberg). Fix $0 < r < 1$. Let φ be a homeomorphism from the unit disc D onto itself such that φ is analytic on $K_r = \{z \in \mathbb{C}: r < |z| < 1\}$ but $\varphi|_{\partial D}$ could not be extended to an analytic function on D. The existence of such a function can be proved by a direct argument but it seems simpler to get it in the following way:
Fix a Jordan curve L contained in $D_r = \{z \in \mathbb{C}: |z| \leq r\}$ which is not the boundary of a disc. Let S be a compact set bounded by L and D. There is (Markuszewicz [1]) an analytic homeomorphism ψ from S onto $K_{r'}$ for some $0 < r' < r$. Let φ be any extension of ψ to a homeomorphism of D onto itself. We define an operator

$$P^+: A(K_r) \to A(D)$$

by

$$P^+\Big(\sum_{n=-\infty}^{+\infty} a_n z^n\Big) = \sum_{n=0}^{+\infty} a_n z^n$$

and we define

$$T: \quad A(D) \to A(D)$$

by

$$T(f) = P^+(F \circ \varphi).$$

Let $f \in A(D)$ and let

$$f \circ \varphi(z) = \sum_{n=-\infty}^{+\infty} a_n z^n \qquad \text{for} \quad z \in \operatorname{int} K_r.$$

We have

$$\|T(f) - f \circ \varphi\|_{\partial D} = \|\sum_{-\infty}^{-1} a_n z^n\|_{\partial D} \leq r\|\sum_{-\infty}^{-1} a_n z^n\|_{|z|=r} \leq \tag{66}$$

$$\leq r\|P^-\|\cdot\|f \circ \varphi\|_{K_r} \leq r\|P^-\|\cdot\|f\|_D,$$

where

$$P^-: A(K_r) \to A(\mathbb{C} \setminus \operatorname{int} D_r): P^-(f) = f - P^+(f).$$

Hence

$$(1 - \|P^-\|\cdot r)\|f\| \leq \|T(f)\| \leq (1 + \|P^-\| \cdot r)\|f\|.$$

By (Voskanian [1]) we have $\lim_{r\to 0} \|P^+\| = 1$, so $\limsup_{r\to 0} \|P^-\| \leq 2$ and we get

$$\lim_{r\to 0} \|T\|\,\|T^{-1}\| = 1 .$$

To end the proof let R be any automorphism of disc algebra, then there exists a Blaschke factor $b: D \to D$ such that $R(f) = f \circ b$ for $f \in A(D)$. Hence, from (66) we get

$$\|f \circ \varphi - f \circ b\| \leq (\|T - R\| + r\|P^-\|)\|f\| \qquad \text{for} \quad f \in A(D). \tag{67}$$

By the definition of φ we have $\varphi|_{\partial D} \neq b|_{\partial D}$, so there is a $z_o \in \partial D$ such that $\varphi(z_o) \neq b(z_o)$. Let $f \in A(D)$ be such that $1 = \|f\| = f(\varphi(z_o)) = -f(b(z_o))$. Hence, by (67) we get

$$\|T - R\| \geq 2 - r\|P^-\| \xrightarrow[r\to 0]{} 2. \tag{68}$$

●

We are ready now to prove that Problem 17.1 has negative solution.

<u>17.5. Definition</u>. Let A be a function algebra. We put

$\varepsilon(A) = \sup\{\varepsilon \geq 0$: for any function algebra B if there is a linear isomorphism $T: A \to B$ with $T(1) = 1$ and $\|T\|\,\|T^{-1}\| \leq 1 + \varepsilon$ then the algebras A and B are isomorphic$\}$,

$\varepsilon_s(A) = \sup \{\varepsilon \geq 0:$ for any function algebra B and any linear isomorphism $T: A \to B$ with $T(\mathbb{1}) = \mathbb{1}$ and $\|T\|\|T^{-1}\| \leq$ $\leq 1 + \varepsilon$ there is an algebra isomorphism $\tilde{T}$ from A onto B such that $\|\tilde{T} - T\| \leq 2 - \varepsilon\}$.

•

17.6. THEOREM. There is a function Dirichlet algebra A such that

$$\inf \{\varepsilon(A|_S): \ S \text{ - maximal set of antisymmetry of } A\} > 0$$

but A is not stable.

Proof. Let D be the unit disc in the complex plane $\mathbb{C}$, $(z_n)_{n=1}^{\infty}$ a dense subset of ∂D and let $(K_n)_{n=1}^{\infty}$ be a sequence of pairwise non-homeomorphic and pairwise disjoint compact subsets of $\mathbb{C}$ such that

$$K_n \cap D = \{z_n\} \qquad \text{for} \quad n = 1,2, \ldots .$$

Put $K = D \cup \bigcup_{n \in \mathbb{N}} K_n$ and put

$$B = \{f \in C(K): \ f|_{\text{int } D} \text{ is holomorphic}\}.$$

The maximal sets of antisymmetry of B are disc and one point sets. B is evidently a Dirichlet algebra and, since the disc algebra is stable (Theorem 18.8), the family

$$\{B|_S: \ S \text{ - maximal set of antisymmetry of } B\}$$

is a uniformly stable family of function algebras.

To prove that B is not stable fix $\varepsilon > 0$. By the proof of Theorem 17.4 - the formula (68) - there is an isomorphism T from the disc algebra onto itself such that $T(\mathbb{1}) = \mathbb{1}$, $\|T\|\cdot\|T^{-1}\| \leq 1 + \varepsilon$ and $\|T - R\| \geq 1$ for any isometry R of $A(D)$. By Theorem 3.1 there is a homeomorphism $\varphi: \partial A \to \partial A$ such that

$$|T(f)(z) - f \circ \varphi(z)| \leq \varepsilon'(\varepsilon)\|f\| \qquad \text{for } f \in A(D), \ z \in D. \tag{69}$$

Let $\tilde{\varphi}$ be any extension of the homeomorphism $\varphi: \partial D \to \partial D$ to a homeomorphism $\tilde{\varphi}$ from K onto a compact subset $\tilde{K}$ of the complex plane which maps D onto D. We put

$$\tilde{B} = \{f \in C(\tilde{K}) : \ f|_{\mathrm{int}\, D} \ \text{is holomorphic}\}.$$

We extend T to an isomorphism $\tilde{T}$ from B onto $\tilde{B}$ putting for $f \in B$

$$\tilde{T}(f)(z) = \begin{cases} T(f|_D)(z) & \text{for} \quad z \in D \\ f(z) + T(f|_D)(z_n) - f(z_n) & \text{for} \quad z \in K_n . \end{cases}$$

From (69) we have

$$\|T\|\,\|T^{-1}\| \le 1 + 3\varepsilon'.$$

Assume there is an isometry S from B onto $\tilde{B}$, then there is an onto homeomorphism $\psi : \tilde{K} \to K$ such that

$$S(f) = f \circ \psi \qquad \text{for} \quad f \in B.$$

Isometry S maps the disc algebra $A(D)$ contained in B onto the corresponding disc algebra in $\tilde{B}$ so $\psi|_D$ is a Blaschke factor. Homeomorphism ψ maps subsets of $\tilde{K}$ onto homeomorphic subsets of K so, since the sets K_n, $n = 1, 2, \ldots$ are pairwise non-homeomorphic, we get $\psi(z_n) = \tilde{\varphi}^{-1}(z_n)$ for any $n \in \mathbb{N}$. Hence

$$\psi|_{\partial D} = \varphi^{-1}|_{\partial D}.$$

This proves that φ is a Blaschke factor and from (69) we have that there is an isometry

$$\Phi : A(D) \to A(D) : \qquad f \mapsto f \circ \varphi$$

with $\|T - \Phi\| \le \varepsilon'$ which contradicts our assumption (if $\varepsilon' < 1$).

∎

In contradistinction to the above theorem the situation is much "nicer" for strongly stable algebras, at least in the separable case.

17.7. Proposition. For any function algebra A we have $\varepsilon_s(A) > 0$ if and only if A is strongly stable.

<u>Proof</u>. The "if" part is evident. To prove the converse implication fix $0 < \varepsilon < \varepsilon_s(A)$. We can assume, taking ε smaller if necessary, that $\varepsilon \leq \varepsilon_0$ and $\varepsilon' < \varepsilon_s(A)$ where ε_0 and ε' are as in Theorem 16.7.

Let T be a linear isomorphism from A onto a function algebra B such that $T(\mathbb{1}) = \mathbb{1}$ and $\|T\|\cdot\|T^{-1}\| \leq 1 + \varepsilon$. By the definition of $\varepsilon_s(A)$ there is an algebra isomorphism $\tilde{T}: A \to B$ such that $\|\tilde{T} - T\| < 2 - \varepsilon'$.

By Theorem 16.7 there is an onto homeomorphism φ: ChB $\to$ ChA such that

$$\|T(f) - f \circ \varphi\| \leq \varepsilon' \|f\| \qquad \text{for } f \in A. \tag{70}$$

On the other hand, since $\tilde{T}$ is an algebra isomorphism, $\tilde{T}$ is of the form

$$\tilde{T}(f) = f \circ \psi \qquad \text{for } f \in A \tag{71}$$

where ψ is a homeomorphism from ChB onto ChA.

From (70),(71) and the inequality $\|\tilde{T} - T\| < 2 - \varepsilon'$ we get

$$|f \circ \varphi(s) - f \circ \psi(s)| < 2\|f\| \qquad \text{for } f \in A,\ s \in \text{ChB}.$$

Hence, as in the proof of Theorem 17.4 we get $\varphi = \psi$ and by (70) and (71) this proves

$$\|\tilde{T} - T\| \leq \varepsilon'.$$

and ends the proof.

∎

<u>17.8. THEOREM</u>. Let A be a function algebra and assume that any maximal set of antisymmetry of A is metrizable. Then if $\delta(A) > 0$, then A is strongly stable, where

$$\delta(A) = \{\inf\ \varepsilon_s(A|_S):\ S - \text{maximal set of antisymmetry of } A\}.$$

<u>Proof</u>. Fix a positive ε such that $\varepsilon < \varepsilon_0$ and $5\varepsilon' \leq \delta(A)$, where ε_0 and $\varepsilon' = \varepsilon'(\varepsilon)$ are as in Theorem 16.7. Let T be a linear isomorphism from A onto a function algebra B with $T(\mathbb{1}) = \mathbb{1}$ and $\|T\|\cdot\|T^{-1}\| \leq 1 + \varepsilon$ and let φ be a homeomorphism from ChB onto

ChA such that

$$\|\tilde{T}(f) - T(f)\| \le \varepsilon'\|f\| \qquad \text{for} \qquad f \in A \tag{72}$$

where the map $\tilde{T}\colon A \to C(\mathrm{ChB})$ is defined by $\tilde{T}(f) = f \circ \varphi$.

We shall prove that $\tilde{T}$ is an algebra isomorphism from A onto $B \subset C(\mathrm{ChB})$.

Fix a maximal set of antisymmetry $S_o \subset \partial B$ of B and let

$$T_1\colon B|_{S_o} \to A|_{cl(\varphi(S_o \cap \mathrm{ChB}))}$$

be defined as in the proof of Theorem 16.7 d). We have

$$\|T_1(g) \circ \varphi - g\| \le 3\varepsilon'\|g\| \qquad \text{for} \qquad g \in B|_{S_o} \tag{73}$$

and

$$\|T_1\|\cdot\|T_1^{-1}\| \le 1 + 5\varepsilon', \qquad T_1(\mathbb{1}) = \mathbb{1}.$$

By Theorem 16.7 c), $cl(\varphi(S_o \cap \mathrm{ChB}))$ is a maximal set of antisymmetry of A so, by our assumption, there is an algebra isomorphism $\tilde{T}_1\colon B|_{S_o} \to A|_{cl(\varphi(S_o \cap \mathrm{ChB}))}$ such that

$$\|\tilde{T}_1 - T_1\| \le 2 - 5\varepsilon'. \tag{74}$$

As before we get that there is a homeomorphism ψ from $\mathrm{Ch}(A|_{cl(\varphi(S_o \cap \mathrm{ChB}))}) = \varphi(S_o \cap \mathrm{ChB})$ onto $\mathrm{Ch}(B|_{S_o}) = S_o \cap \mathrm{ChB}$ such that

$$\tilde{T}_1(g) = g \circ \psi \qquad \text{for} \qquad g \in B|_{S_o}.$$

From (73),(74) we find

$$\|g \circ \psi - g \circ \varphi^{-1}\| \le 2 - 5\varepsilon' + 3\varepsilon' \qquad \text{for} \qquad g \in B|_{S_o}$$

and hence $\psi = \varphi^{-1}$.

The above proves that for any $f \in A$ we have

$$f \circ \varphi|_{S_o} \in B|_{S_o}.$$

Since S_o is an arbitrary maximal set of antisymmetry of B, by the Shilov-Bishop theorem this proves that

$$\tilde{T}(A) \subseteq B \subseteq C(\mathrm{ChB}).$$

Hence, since T is an onto map, (72) proves that $\tilde{T}$ is also onto and this means that it is an algebra isomorphism from A onto B so the theorem is proved.

•

§ 18. Deformations of algebras of functions on Riemann surfaces.

One of the most important classes of function algebras are algebras of analytic functions defined on a one dimensional Riemann surface. While the general theory of perturbations of function algebras is only in the initial stages of developments, Rochberg's paper [7] gives almost a complete description of small perturbations in the class of algebras of analytic functions defined on a one dimensional Riemann surface. In this section we present the Rochberg's results.

First we introduce a bit more notation and definitions:

- $\mathfrak{S}$ - the set of all connected finite bordered one dimensional Riemann surfaces;
- $A(S)$, for $S \in \mathfrak{S}$ - the uniform algebra of functions continuous on S and analytic at the non-boundary points of S;
- $A(\mathfrak{S}) = \{A(S): S \in \mathfrak{S}\}$;
- ∂S, for $S \in \mathfrak{S}$ - the border of S;
- $\mathfrak{A} = \{A \in \mathcal{A}: A$ is a finite codimensional subalgebra of finite direct sums of elements from $A(\mathfrak{S})$ such that $\mathfrak{M}(A)$ is connected, $\partial A = \mathrm{Ch} A$ and ∂A is homeomorphic to a finite union of circles$\}$;
- $X(A) = C_R(\partial A)/\overline{\mathrm{Re} A}$, for $A \in \mathcal{A}$;
- $k(A) = \dim (X(A))$, for $A \in \mathcal{A}$;
- $\mathfrak{A}_k = \{A \in \mathfrak{A}: k(A) = k\}$;
- $\mathfrak{S}^*$ - the set of finite bordered connected possibly singular surfaces which are obtained as $\mathfrak{M}(A)$ for A in $\mathfrak{A}$;
- $A(S)$, for $S \in \mathfrak{S}^*$ - the algebra of continuous functions on S holomorphic on $S \setminus \partial S$.

The main results concern the set $\mathfrak{A}$ regarded as a subset of $\mathcal{A}$. We omit the proofs, which are not easy, and can be found in Rochberg's paper [7].

We first recall Sawoń & Warzecha [1] and Warzecha [1] results which show how the algebras in $\mathfrak{A}_k$ looks like.

18.1. Theorem (Sawoń and Warzecha). Let A be a Banach algebra. We have

a) if A' is a finite codimensional subalgebra of A then there is a sequence $A' = A_n \subset A_{n-1} \subset \dots \subset A_1 \subset A_o = A$ of function algebras such that

$$\dim ({}^{A}k-1/A_k) = 1 \qquad \text{for} \quad k = 1,2, \dots ,n;$$

b) if A_1 is a codimension one subalgebra of A then one of following three possibilities holds:

1. $A_1 = \ker F$ for some $F \in \mathfrak{M}(A)$,
2. $A_1 = \ker D$ for some point derivation D defined on A,
3. $A_1 = \ker (F_1 - F_2)$ for some $F_1, F_2 \in \mathfrak{M}(A)$.

•

Where by a point derivation D defined on a Banach algebra A we mean, as usual, a functional on A such that

$$D(fg) = F(f)D(g) + D(f)F(g) \qquad \text{for} \quad f,g \in A$$

with F a fixed element of $\mathfrak{M}(A)$.

18.2. THEOREM (Rochberg). Suppose A_1, A_2 are in $\mathfrak{A}$ and assume $d(A_1,A_2) = 0$ then $A_1 = A_2$.

18.3. THEOREM (Rochberg). For any non-negative integer k there are positive constants ε_k and c_k such that for any $A \in \mathfrak{A}_k$ we have: If B is in $\mathcal{A}$ and if T is a linear isomorphism from A onto B with $T(\mathbb{1}) = \mathbb{1}$ and $\|T\|\cdot\|T^{-1}\| \le 1 + \varepsilon \le 1 + \varepsilon_k$ then there is an S in $\mathfrak{S}$ such that B is a subalgebra of $A(S)$, $\partial B = \partial S$, $\mathfrak{M}(B) = S$. Furthermore there is a quasiconformal homeomorphism τ of a neighbourhood U_B of ∂S onto a neighbourhood U_A of ∂A, with

$$\sup_{U_B \setminus \partial S} |\text{dilitation of } \tau| \le 1 + c_k\varepsilon$$

and such that

$$|Tf(s) - f \circ \tau(s)| \leq c_k \varepsilon \|f\| \quad \text{for} \quad f \in A, \quad s \in U_B.$$

Note that in the above theorem it is not claimed that B is in $\mathfrak{A}$. That is, it is not claimed that B is of finite codimension in $A(S)$. The theorems which follow show that this is true if further assumption are made on B or if we allow the constant ε_k to depend on A.

18.4. THEOREM (Rochberg). Each $\mathfrak{A}_k$ is a closed and open subset of $\mathcal{A}$.

The following theorems give more informations about the types of changes which may occur as we pass from $A \in \mathfrak{A}_k$ to a nearby B.

Recall that for $S, S' \in \mathfrak{S}$, the Teichmüller distance between S and S' is defined by

$$d_T(S,S') = \inf\{\log K\text{: there is a } K\text{-quasiconformal homeomorphism of } S \text{ onto } S'\}.$$

18.5. THEOREM (Rochberg). Let $A(S) \in A(\mathfrak{S})$. There is an $\varepsilon = \varepsilon(A) > 0$ such that the set of function algebras B with $d(A,B) < \varepsilon$ consists entirely of algebras of the form $B = A(S')$ for some $S' \in \mathfrak{S}$. Furthermore, for such B, $d(A,B) \sim d_T(S,S')$. Also, if S'' is any element of $\mathfrak{S}$ with $d_T(S,S'')$ sufficiently small then $d(A,A(S'')) < \varepsilon$.

For example, as an immediate consequence, we get the following corollary to the above Theorem.

18.6. Corollary. Let $K_r = \{z \in \mathbb{C}\colon \ r \leq |z| \leq 1\}$ for some r, $0 < r < 1$. We have $K_r \in \mathfrak{S}$, $A(K_r) \in \mathfrak{A}_1$. There is an $\varepsilon(r) > 0$ such that for any $B \in \mathcal{A}$ with $d(A(K_r),B) < \varepsilon(r)$ we have $B = A(K_t)$ with $|t - r| \leq O(\varepsilon)$.

18.7. THEOREM (Rochberg). Let A be in $\mathfrak{A}_k$. There is a finite set R in $\mathfrak{M}(A)$ with $\text{Sing}\,(A) \subset R$ such that given any open neighbourhood N of R there are constants ε_0 and c so that the following holds. If $B \in \mathcal{A}$ and T is a linear isomorphism from A onto B with $T\mathbb{1} = \mathbb{1}$ and $\|T\|\cdot\|T^{-1}\| \leq 1 + \varepsilon$ then there is an open set U in $\mathfrak{M}(B)$, $\partial B \subseteq U$ and a homeomorphism τ of U onto $\mathfrak{M}(A) \setminus N$ so that

$$|Tf(x) - f \circ \tau(x)| \leq c\varepsilon\|f\| \quad \text{for all } f \text{ in } A, \ x \text{ in } U$$

and

$$\sup_{U \setminus \partial B} |\text{dilitation of } \tau| \leq 1 + c\varepsilon.$$

It is not difficult to deduce from the above theorems that the disc algebra is an unique stable algebra in $\mathfrak{A}$.

18.8. THEOREM (Rochberg). Let $A \in \mathfrak{A}$ then A is stable if and only if A is a disc algebra.

§ 19. The Hochschild cohomology groups and small perturbations.

In 1977 Johnson [3] proved that the perturbation theory for general Banach algebras is closely related to questions involving Banach algebra cohomology.

In this section we give some of his results.

We first recall the definition of the Hochschild cohomology groups.

For a Banach algebra A and a positive integer n we denote by $\mathcal{L}^n(A,A)$ the Banach space of all n-linear continuous forms from $A\times \dots \times A$, the product of n copies of A, into A. We define a sequence $(\delta_n)_{n=1}^{\infty}$ of linear maps

$$\xrightarrow{\delta_{n-1}} \mathcal{L}^n(A,A) \xrightarrow{\delta_n} \mathcal{L}^{n+1}(A,A) \xrightarrow{\delta_{n+1}}$$

by

$$\delta_n(T)(a_1, \ldots, a_n, a_{n+1}) =$$

$$= a_1 \cdot T(a_2, \ldots, a_{n+1}) + \sum_{j=1}^{n} (-1)^j T(a_1, \ldots, a_j \cdot a_{j+1}, \ldots, a_{n+1}) +$$

$$+ (-1)^{n+1} T(a_1, \ldots, a_n) \cdot a_{n+1}.$$

We put

$$\left.\begin{aligned} \mathcal{Z}^n(A,A) &= \ker \delta_n \\ \mathcal{N}^n(A,A) &= \operatorname{Im} \delta_{n-1} \end{aligned}\right\} \quad \text{for} \quad n \in \mathbb{N}.$$

By a direct calculation we get $\mathcal{N}^n(A,A) \subseteq \mathcal{Z}^n(A,A)$ and we define the n-th Hochschild cohomology group of A by

$$\kappa^n(A,A) = \mathcal{Z}^n(A,A) / \mathcal{N}^n(A,A) \qquad \text{for} \quad n \in \mathbb{N}.$$

For general cohomological results in Banach algebras see Johnson [1].

<u>19.1. THEOREM (Johnson)</u>. Let A be a Banach algebra with $\kappa^2(A,A) = 0$ and assume $\mathcal{N}^3(A,A)$ is closed in $\mathcal{L}^3(A,A)$. Then A is strongly stable.

<u>Proof</u>. By the open mapping theorem there are positive numbers K and L such that:

for any $T \in \mathcal{Z}^2(A,A)$ there is an $S \in \mathcal{L}(A,A)$ with $\delta_1 S = T$ and $\|S\| \leq K\|T\|$

and

for any $T \in \mathcal{N}^3(A,A)$ there is an $S \in \mathcal{L}^2(A,A)$ with $\delta_2 S = T$ and $\|S\| \leq L\|T\|$.

Put

$$q(x) = K(x + 2Lx^2)$$

and

$$p(x) = (1 - q(x))^{-2}(2Lx^2 + xq(x) + (q(x))^2), \quad \text{for } x \in \mathbb{R}^+.$$

We have $\lim_{x\to 0} \frac{p(x)}{x} = 0$, so there is an $\varepsilon_o > 0$ such that

$$p(x) < x \quad \text{for} \quad x \in (0,\varepsilon_o) \quad \text{and} \quad q(\varepsilon_o) < 1.$$

Let $\times$ be a new multiplication on A with

$$\|\times - \cdot\| = \varepsilon_1 \leq \varepsilon_o.$$

We shall prove that there is a linear continuous map $T: A \to A$ such that

$$a \times b = T^{-1}(Ta \cdot Tb) \quad \text{for} \quad a,b \in A$$

and

$$\|T - Id\| \leq \varepsilon'(\varepsilon_1) \quad \text{where} \quad \varepsilon'(\varepsilon) \to 0^+ \quad \text{as} \quad \varepsilon \to 0^+.$$

For all a,b,c in A we have

$$\delta_2(\times - \cdot)(a,b,c) = \delta_2(\times)(a,b,c) = a\cdot(b \times c) - (ab) \times c + a \times (bc) +$$

$$- (a \times b)\cdot c = a\cdot(b \times c - bc) + (ab - a \times b)\cdot c - a \times (b \times c - b\cdot c) +$$

$$- (ab - a \times b) \times c$$

so that

$$\|\delta_2(\times - \cdot)(a,b,c)\| \leq \varepsilon\|a\|\cdot\|b \times c - bc\| + \varepsilon\|ab - a \times b\|\cdot\|c\| \leq$$

$$- 2\varepsilon^2\|a\|\cdot\|b\|\cdot\|c\|.$$

Hence there is a $T_1 \in \mathcal{L}^2(A,A)$ with $\|T_1\| \leq 2L\varepsilon_1^2$ and $\delta_2(T_1) = \delta(\times - \cdot)$ and there is an $S_1 \in \mathcal{L}(A,A)$ with $\delta_1(S_1) = \times - \cdot - T_1$ and $\|S_1\| \leq K\|\times - \cdot - T_1\| \leq q(\varepsilon_1) < 1$. We define the third multiplication $\times_1$ on A by

$$a \times_1 b = (Id + S_1)((Id + S_1)^{-1}(a) \times (Id + S_1)^{-1}(b)) \quad \text{for all } a,b \text{ in } A.$$

Fix $a,b \in A$ and put $a' = (Id + S_1)^{-1}(a)$, $b' = (Id + S_1)^{-1}(b)$. We have

$$ab - a \times_1 b = (Id + S_1)(a')\cdot(Id + S_1)(b') - (Id + S_1)(a' \times b') =$$

$$= a'b' + \delta_1(S_1)(a',b') + S_1(a'\cdot b') + S_1(a')S_1(b') - a' \times b' +$$

$$- S_1(a' \times b') = T_1(a',b') - S_1(a' \times b' - a'\cdot b') + S_1(a')S_1(b')$$

hence

$$\|\cdot - \times_1\| \leq (1 - \|S_1\|)^{-2}(\|T_1\| + \varepsilon_1\|S_1\| + \|S_1\|^2) \leq p(\varepsilon_1) < \varepsilon_1.$$

By repeating the process we obtain a sequence $S_1, S_2, \ldots$ in $\mathcal{L}(A)$, a sequence $\times_1, \times_2, \ldots$ of multiplications on A and a sequence $\varepsilon_1, \varepsilon_2, \ldots$ of positive numbers. We have $p(x) = O(x^2)$ as $x \to 0$ so the series $\Sigma\varepsilon_i$ converges hence $\Sigma\|S_i\|$ converges. Thus the product $T_n = (Id + S_n)\circ(Id + S_{n-1})\circ \ldots \circ(Id + S_1)$ converges as $n \to \infty$ to an invertible element T_∞ of the algebra $\mathcal{L}(A)$ and

$$\|T_\infty\|\cdot\|T_\infty^{-1}\| \leq 1 + O(\varepsilon_1).$$

Furthermore

$$a \times b = T_\infty^{-1}(T_\infty a \cdot T_\infty b) \quad \text{for any } a,b \text{ in } A.$$

●

We have the following examples of Banach algebras for which the hypotheses of Theorem 19.1 are satisfied.

a) $A = K(H)$, with H a Hilbert space (Johnson [2]).

b) $A = \mathcal{L}(X)$, with X a Banach space (Kaliman and Selivanov [1]).

c) $A = L^1(G)$, with G amenable group (Johnson [2]).

d) $A = C(S)$, with S metric compact set (Johnson [3])(1).

Let H be a Hilbert space. From Corollary 13.3 we get that there is an $\varepsilon_0 > 0$ such that for any ε-isometry T from $K(H) = H^* \check{\otimes} H$ onto itself with $\varepsilon < \varepsilon_0$ we have $\|T - T_1 \otimes T_2\| = O(\varepsilon)$ where T_1, T_2 are $O(\varepsilon)$-isometries. On the other hand it is well-known that any algebra isomorphism T form $K(H)$ onto itself is of the form $T = T_1 \otimes T_2$ where $T_1\colon H^* \to H^*$ and $T_2\colon H \to H$ are onto linear isomorphisms. Hence by a) above and Johnson's Theorem we get a partial conversion of Corollary 13.3.

(1) The author does not know any other example of function algebra which satisfies the assumptions of Theorem 19.1.

19.2. Corollary. Let T be an ε-algebra isomorphism from $K(H) = H^* \check{\otimes} H$ onto itself then there are linear onto isomorphisms $T_1: H^* \to H^*$ and $T_2: H \to H$ such that

$$\|T - T_1 \otimes T_2\| = O(\varepsilon).$$

§ 20. Perturbations of topological algebras.

In this section we generalize the idea of small perturbation into topological algebras. We shall prove that Theorem 3.1 can be extended to topological function algebras. It shows, roughly speaking, that if topological function algebra A is an inverse limit of a system of function algebras $\{A_\alpha : \alpha \in \Lambda\}$ then the investigation of small perturbations of A can be restricted to the investigation of small perturbations of A_α; $\alpha \in \Lambda$.

20.1. Definition. By a topological algebra we mean a triple $(A, \cdot, \{\|\cdot\|_\alpha ; \alpha \in \Lambda\})$ where $(A, \cdot)$ is an algebra with unit and $(A, \{\|\cdot\|_\alpha : \alpha \in \Lambda\})$ is a complete topological vector space. We assume also that the norms $\|\cdot\|_\alpha$, $\alpha \in \Lambda$ are sumbultiplicative, this means that

$$\|a \cdot b\|_\alpha \leq \|a\|_\alpha \|b\|_\alpha \quad \text{for all} \quad \alpha \in \Lambda; \quad a, b \in A.$$

•

For a topological algebra $(A, \cdot, \{\|\cdot\|_\alpha : \alpha \in \Lambda\})$ we denote by $\tilde{A}_\alpha$ the algebra obtained from A by identifying elements from $\ker \|\cdot\|_\alpha$ with zero. We denote by A_α the completion of $\tilde{A}_\alpha$ in the norm given by $\|\cdot\|_\alpha$. $(A_\alpha, \alpha \in \Lambda)$ is an inverse system of Banach algebras and $A = \varprojlim A_\alpha$.

20.2. Definition. A topological algebra $(A, \cdot, \{\|\cdot\|_\alpha : \alpha \in \Lambda\})$ is called topological function algebra if $(A_\alpha, \|\cdot\|_\alpha)$ is a function algebra for any $\alpha \in \Lambda$.

•

20.3. Definition. Let S be a topological Hausdorff space. We denote by $\tilde{C}(S)$ the algebra of all continuous scalar-valued functions defined on S.

•

20.4. Example. Let S be topological vector space and let $\{S_\alpha : \alpha \in \Lambda\}$ be a family of compact subsets of S such that $\bigcup\{S_\alpha : \alpha \in \Lambda\} = S$. We define seminorms $\|\cdot\|_\alpha$ for $\alpha \in \Lambda$ by

$$\|f\|_\alpha = \sup\{|f(s)| : s \in S_\alpha\} \quad \text{for } f \in \tilde{C}(S), \quad \alpha \in \Lambda.$$

The triple $(\tilde{C}(S), \cdot, \{\|\cdot\|_\alpha : \alpha \in \Lambda\})$ is a topological function algebra; here $\cdot$ is a usual point-wise multiplication.

•

The following is a well-known fact.

20.5. Proposition. Let $(A, \cdot, \{\|\cdot\|_\alpha : \alpha \in \Lambda\})$ be a topological function algebra. There is a topological space S, a family $\{S_\alpha : \alpha \in \Lambda\}$ of compact subsets of S and an algebra isomorphism T between A and a closed subalgebra B of $(\tilde{C}(S); \{S_\alpha$ compact subset of S, $\alpha \in \Lambda\})$ such that

$$\|f\|_\alpha = \sup\{|Tf(s)| : s \in S_\alpha\} \quad \text{for } f \in A, \quad \alpha \in \Lambda.$$

•

In the sequel we shall identify, via the above Proposition, any topological function algebra with a closed subalgebra of $(\tilde{C}(S), \{S_\alpha$ compact subset of S, $\alpha \in \Lambda\})$, for some topological space S.

20.6. THEOREM. Let $(A, \cdot, \{\|\cdot\|_\alpha : \alpha \in \Lambda\})$ be a complex topological function algebra and let $\times_n$, $n \in \mathbb{N}$ be a sequence of multiplications defined on the vector space A. Assume that multiplications $\times_n$, $n \in \mathbb{N}$ have the same unit as the original multiplication on A. Then the following are equivalent.

(i) there is a sequence of positive numbers $(\varepsilon_n)_{n=1}^\infty$ with $\lim_n \varepsilon_n = 0$ such that

$$\big|\,\|f \cdot g\|_\alpha - \|f \times_n g\|_\alpha\big| \le \varepsilon_n \|f\|_\alpha \|g\|_\alpha \quad \text{for any } \alpha \in \Lambda, \quad n \in \mathbb{N}, \quad f, g \in A;$$

(ii) there is a sequence of positive numbers $(\varepsilon_n)_{n=1}^{\infty}$ with $\lim_n \varepsilon_n = 0$ such that

$$\|f \cdot g - f \times_n g\|_\alpha \leq \varepsilon_n \|f\|_\alpha \|g\|_\alpha \qquad \text{for any } \alpha \in \Lambda, \quad n \in \mathbb{N}, \quad f,g \in A;$$

(iii) there is a sequence of positive numbers $(\varepsilon_n)_{n=1}^{\infty}$ with $\lim_n \varepsilon_n = 0$ such that

$$\|f \times_n g\|_\alpha \leq (1 + \varepsilon_n)\|f\|_\alpha \|g\|_\alpha \qquad \text{for any } \alpha \in \Lambda, \quad n \in \mathbb{N}, \quad f,g \in A;$$

(iv) there is a sequence of positive numbers $(\varepsilon_n)_{n=1}^{\infty}$ with $\lim_n \varepsilon_n = 0$, a sequence of topological algebras $(B_n, \cdot, \{\|\cdot\|_\alpha, \alpha \in \Lambda\})$ and a sequence of linear onto isomorphisms $T_n: A \to B_n$ such that

$$f \times_n g = T_n^{-1}(T_n f \cdot T_n g) \qquad \text{for} \quad n \in \mathbb{N}, \quad f,g \in A$$

and

$$(1 - \varepsilon_n)\|f\|_\alpha \leq \|T_n(f)\|_\alpha \leq (1 + \varepsilon_n)\|f\|_\alpha \qquad \text{for} \quad \alpha \in \Lambda, \quad n \in \mathbb{N}, \quad f \in A.$$

Proof. The implications (ii) ⇒ (i) ⇒ (iii) and (iv) ⇒ (iii) are trivial.

Assume that (iii) is fulfilled, we get

$$\|f\|_\alpha = 0 \Rightarrow \|f \times_n g\|_\alpha = 0 \qquad \text{for any} \quad \alpha \in \Lambda, \quad n \in \mathbb{N}, \quad g \in A.$$

Hence the multiplication $\times_n$ define a multiplication on a function algebra $(A_\alpha, \|\cdot\|_\alpha)$; we denote this new multiplication on $(A_\alpha, \|\cdot\|_\alpha)$ also by $\times_n$.

The above observations and Theorem 3.1 gives the implication (iii) ⇒ (ii). We get also that for any $\alpha \in \Lambda$ there is a sequence $(B_n^\alpha)_{n=1}^{\infty}$ of function algebras and a sequence $(T_n^\alpha)_{n=1}^{\infty}$ of linear isomorphisms from A_α onto B_n^∞ such that

$$f \times_n g = (T_n^\alpha)^{-1}(T_n^\alpha f \cdot T_n^\alpha g) \qquad \text{for} \quad \alpha \in \Lambda, \quad n \in \mathbb{N}, \quad f,g \in A_\alpha \tag{75}$$

and

$$\lim_n \sup_\alpha \|T_n^\alpha\| = 1 = \lim_n \sup_\alpha \|(T_n^\alpha)^{-1}\|. \tag{76}$$

We define an order on Λ by

$$\alpha < \alpha' \equiv \|\cdot\|_\alpha \leq \|\cdot\|_{\alpha'} .$$

For any $\alpha < \alpha'$ from Λ we put

$$S_n^{\alpha',\alpha} = T_n^\alpha \circ \tau_{\alpha',\alpha} \circ (T_n^{\alpha'})^{-1} : B_n^{\alpha'} \to B_n^\alpha ;$$

here $\tau_{\alpha',\alpha}$ denotes the natural algebra homomorphism from $A_{\alpha'}$ into A_α.

For any $f,g \in B_n^{\alpha'}$ we have

$$S_n^{\alpha,\alpha'}(f \cdot g) = T_n^\alpha \circ \tau_{\alpha,\alpha'}((T_n^{\alpha'})^{-1}(f) \times_n (T_n^{\alpha'})^{-1}(g)) =$$

$$= T_n^\alpha(\tau_{\alpha,\alpha'}((T_n^{\alpha'})^{-1}(f)) \times_n \tau_{\alpha,\alpha'}((T_n^{\alpha'})^{-1}(g))) = S_n^{\alpha,\alpha'}(f) \cdot S_n^{\alpha,\alpha'}(g) .$$

So $S_n^{\alpha,\alpha'}$ is an algebra homomorphism from $B_n^{\alpha'}$ into B_n^α.

For any $n \in \mathbb{N}$ we have the following commutative diagram, for any $\alpha < \alpha' < \alpha''$ from Λ:

$$\begin{array}{ccccccccc}
\longrightarrow & A^{\alpha''} & \xrightarrow{\tau_{\alpha'',\alpha'}} & A^{\alpha'} & \xrightarrow{\tau_{\alpha',\alpha}} & A^{\alpha} & \longrightarrow \\
 & \downarrow T_n^{\alpha''} & & \downarrow T_n^{\alpha'} & & \downarrow T_n^{\alpha} & \\
\longrightarrow & B_n^{\alpha''} & \longrightarrow & B_n^{\alpha'} & \longrightarrow & B_n^{\alpha} & \longrightarrow
\end{array}$$

Since $S_n^{\alpha,\alpha'}$ are algebra homomorphisms, then it is standard, by (76), to show that the above diagram defines a linear isomorphism T_n from $A = \varprojlim A_\alpha$ onto a function algebra $B = \varprojlim (B_n^\alpha;\ S_n^{\alpha',\alpha} : \alpha,\alpha' \in \Lambda)$ such that

$$\lim_n \sup_\alpha \|T_n^\alpha\|_\alpha = 1 = \lim_n \sup_\alpha \|(T_n)^{-1}\|_\alpha .$$

The above ends the proof of the implication (iii) ⇒ (iv). •

We get a generalization of Theorem 3.1' from function algebras into topological function algebras. Exactly in the same manner one can get also an extension of Theorem 3.1.

Remarks. The results of this chapter are mostly based on the papers of R. Rochberg [4,7], B. E. Johnson [2,3] and the author [7,10].

The results of § 15 were proved by R. Rochberg [5] with the additional assumptions "$\partial A = ChA$" and "any point of ∂A is G_δ", but by Theorem 3.1 we adopt Rochberg's proofs in the general case. Theorem 16.2 is taken from the author's paper [7] and the theorem 16.4 from [10]. The theorem 17.4 which states that the disc algebra is not strongly stable is probably due to B. E. Johnson (unpublished) and the proof of this result was communicated to the author by Richard Rochberg.

All the results of § 18 are taken from Rochberg's paper [7]. The "cohomological" Theorem 19.1 is taken from Johnson's paper [3].

PROBLEMS

1. Give an example of stable non-Dirichlet function algebra.

2. Is there any strongly stable function algebra not of the form $C(S)$, with S compact Hausdorff space?

2'. Is there a function algebra A not of the form $C(S)$ with S compact (metric) space such that

$$H^2(A,A) = 0 = H^3(A,A)?$$

3. Is the property "A is a logmodular function algebra" stable?

4. Let $H^\infty(D)$ be an algebra of all bounded analytic functions defined on the unit disc. Is $H^\infty(D)$ stable?

5. Let A be a Banach algebra (function algebra) and assume F is an ε-multiplicative linear functional on A. Does then exist a linear and multiplicative functional $\tilde{F}$ on A such that $\|F - \tilde{F}\| \leq O(\varepsilon)$?

6. Does there exist a non-stable Banach algebra (function algebra) which has only countably many small perturbations?

7. Let A and B be function algebras. Are the following implications true?
   a) A and B are stable $\Rightarrow A \check{\otimes} B$ is stable;
   b) $A \check{\otimes} B$ is stable $\Rightarrow A$ and B are stable.

8. Let A be a stable function algebra on a compact Hausdorff space X and let $X_o = \bar{X}_o \subset X$ be a weak peak set for A (a maximal set of antisymmetry of A). Is $A|_{X_o}$ stable?

9. Compute $\varepsilon(C(S))$ for compact Hausdorff spaces S.

10. Compute $\varepsilon(A(D))$ where $A(D)$ is the disc algebra.

11. Let A be a function algebra and assume there is an $\varepsilon_o > 0$ such that for any $B \in \mathcal{A}$ we have $d(A,B) < \varepsilon_o \Rightarrow d(A,B) = 0$. Is then A stable?

12. Is the set of stable function algebras closed in $(\mathcal{A}, d)$?

13. For a compact subset K of the complex plane we denote by $R(K)$ the function algebra generated by rational functions with poles off K. Characterize these compact subsets K of the complex plane $\mathbb{C}$ for which $R(K)$ is stable.

14. Put $B = \{(z,w) \in \mathbb{C}^2 : |z|^2 + |w|^2 \leq 1\}$ and $A(B) = \{f \in C(B) : f|_{\text{int } B}$ is holomorphic$\}$. Is $A(B)$ stable?

15. Can we omit the assumption "... is metrizable" in the theorems 16.7 d) and 17.8?

16. Let A be a function algebra, let $\times$ be a small perturbation of A and let φ be as in Theorem 3.1. Assume $\mathrm{Ch}A = \partial A$. Does there exist an extension of φ to a homeomorphism between neighbourhoods of ∂B and ∂A, respectively?

17. Is there a positive ε_0 such that for any compact subsets X,Y of $\mathbb{R}$ we have
$d(C^1(X), C^1(Y)) < \varepsilon \Rightarrow X$ and Y are homeomorphic?
Here we denote by $C^1(X)$ the space of continuously differentiable functions on X with the norm given by

$$\|f\| = \sup_{t\in X} |f(t)| + \sup_{t\in X} |f'(t)|.$$

18. Can we assume in the theorem 12.1 "X is strictly convex"" instead of "X** is strictly convex"?

19. Is any C*-Banach algebra stable?

20. Is the property "a function algebra A has n generators, $n \in \mathbb{N} \cup \{\infty\}$" stable?

References

L. Ahlfors

[1] Lectures on quasiconformal mappings, Van Nostrand, Princeton, New York, 1966.

D. Amir

[1] "On isomorphisms of continuous function spaces", Israel J. Math. 3, 1965, 205-210.

E. Behrends

[1] M-Structure and the Banach-Stone theorem, Lecture Notes in Mathematics 736, Springer-Verlag, Berlin, 1979.

[2] "Multiplier representations and an application to the problem whether $A \times_{\varepsilon} X$ determines A and/or X" Math.Scan.52,1983.

Y. Benyamini

[1] "Small into-isomorphisms between spaces of continuous functions", Proc. Amer. Math. Soc. 83 (3) 1981, 479-485.

M. Cambern

[1] "Isometries of certain Banach algebras", Studia Math. 25, 1965, 217-225.

[2] "On Isomorphisms with a small bound", Proc. Amer. Math. Soc. 18, 1967, 1062-1066.

[3] "Isomorphisms of spaces of continuous vector-valued functions", Ill. J. Math. 20, 1976, 1-11.

M. Cambern and V. Pathak

[1] "Isometries of spaces of differentiable functions", Math. Japonica, 26 (3), 1981, 253-260.

B. Cengiz

[1] "A generalization of the Banach—Stone theorem", Proc. Amer. Math. Soc. 40, 1973, 426-430.

E. Christensen

[1] "Perturbations of type I von Neumann algebras", J. London Math. Soc. 9 (2), 1975, 395-405.

[2] "Perturbations of operator algebras", Invent. Math. 43, 1977, 1-13.

[3] "Derivations and perturbations", Proc. Symp. Pure Math. 38, 1982, Part 2.

H. B. Cohen

[1] "A bound-two isomorphism between $C(X)$ Banach spaces", Proc. Amer. Math. Soc. 50, 1975, 215-217.

H. M. Farkas and I. Kra

[1] Riemann surfaces, Springer-Verlag (Graduate Texts in Math. 71), 1980.

T. W. Gamelin

[1] Uniform algebras, Prentice-Hall, Englewood Cliffs, 1969.

R. A. Hirschfeld and W. Żelazko

[1] "On spectral norm Banach algebras", Bull. Acad. Polon. Sci. 16, 1968, 195-199.

W. Holsztyński
[1] "Continuous mappings induced by isometries of spaces of continuous functions", Studia Math. 26, 1966, 133-136.

K. Jarosz
[1] "A generalization of the Banach-Stone theorem", Studia Math. 73, 1982, 33-39.
[2] "Perturbations of uniform algebras", Bull. London Math. Soc. 15, 1983, 133-138.
[3] "Metric and algebraic perturbations of function algebras", Proc. Edinburgh Math. Soc. 26, 1983, 383-391.
[4] "The uniqueness of multiplication in function algebras", Proc. Amer. Math. Soc. 89 (2), 1983, 249-253.
[5] "Into isomorphisms of spaces of continuous functions", Proc. Amer. Math. Soc. 90 (3), 1984, 373-377.
[6] "Isometries between injective tensor products of Banach spaces", to appear in Pacific J. Math.
[7] "A characterization of weak peak sets for function algebras", Bull. Australian Math. Soc. 30 (1), 1984, 129-135.
[8] "Isometries in semisimple, commutative Banach algebras", to appear in Proc. Amer. Math. Soc.
[9] "Small isomorphisms between operator algebras", to appear Proc. Edinburgh Math.Soc.
[10] "Perturbations of uniform algebras, II", to appear J.London.Math.S.

B. E. Johnson
[1] Cohomology in Banach algebras, Memoirs, A. M. S. 127 (Providence, R.I. 1972).
[2] "Approximate diagonals and cohomology of certain annihilator Banach algebras", Amer. J. Math. 94, 1972, 685-698.
[3] "Perturbations of Banach algebras", Proc. London Math. Soc. 35 (3) 1977, 439-458.
[4] "A counter-example in the perturbation theory of C*-algebras", preprint, University of Newcastle upon Tyne, 1980.

R. V. Kadison and D. Kastler
[1] "Perturbations of von Neumann algebras I, Stability of type", Amer. J. Math. 94, 1972, 38-54.

Sh. I. Kaliman and Yu. V. Selivanov
[1] "On cohomologies of operator algebras", Vestnik Moskov. Univ. Ser. I Mat. Meh. 5, 1974, 24-27.

R. Larsen
[1] Banach algebras, Pure and Applied Math. 24, Marcel Dekker Inc., N. Y., 1973.

A. I. Markushevich
[1] Teoriya analiticheskikh funkcij I, II, Nauka, Moskva, 1968.

M. Nagasawa
[1] "Isomorphisms between commutative Banach algebras with application to rings of analytic functions", Kodai Math. Sem. Rep. 11, 1959, 182-188.

V. Pathak
[1] "Isometries of $C^{(n)}(0,1)$", Pacific J. Math. 94 (1), 1981, 211-222.
[2] "Linear isometries of spaces of absolutely continuous functions", Can. J. Math. 34 (2), 1982, 298-306.

A. Pełczyński
[1] Linear extensions, Linear averagings, and their applications to linear topological classification of spaces of continuous functions, Dissertationes Math. (Rozprawy Mat.) 58, 1968.

J. Philips
[1] "Perturbations of C*-algebras", Indiana Univ. Math. J. 23, 1974, 1167-1176.

J. Philips and I. Raeburn
[1] "Perturbations of AF-algebras", Canad. J. Math. 31, 1979, 1012-1016.
[2] "Perturbations of C*-algebras, II", Proc. London Math. Soc. 43 (3), 1981, 46-72.

I. Raeburn and J. L. Taylor
[1] "Hochschild cohomology and perturbations of Banach algebras", J. Func. Anal. 25, 1977, 258-266.

N. V. Rao and A. K. Roy
[1] "Linear isometries of some function spaces", Pacific J. Math. 38 (1), 1971, 177-192.

R. Rochberg
[1] "Almost isometries of Banach spaces and moduli of planar domains", Pacific J. Math., 49, 1973, 445-466.
[2] "Almost isometries of Banach spaces and moduli of Riemann surfaces", Duke Math. J., 40, 1973, 41-52.
[3] "Almost isometries of Banach spaces and moduli of Riemann surfaces, II", Duke Math. J. 42, 1975, 167-182.
[4] "The Banach-Mazur distance between function algebras on degenerating Riemann surfaces", Lecture Notes in Math. 604, 1977, 82-94.
[5] "Deformation of uniform algebras", Proc. London Math. Soc. 39 (3) 1979, 93-118.
[6] "The disc algebra is rigid", Proc. London Math. Soc. 39 (3) 1979, 119-130.
[7] "Deformation of uniform algebras on Riemann surfaces", manuscript 1982.

W. Rudin
[1] Functional Analysis, McGraw-Hill Book Company, 1973.

Z. Sawoń and A. Warzecha
[1] "On the general form of subalgebras of codimension 1 of B-algebras with a unit", Studia Math. 29, 1968, 249-260.

E. L. Stout
[1] The theory of uniform algebras, Bogden and Quigley, Belmont, California, 1971.

I. Suciu
[1] Function algebras, Noordhoff International Publishing, Leyden, 1975.

V. V. Voskanian
[1] "Ob odnoj proektsionnoj postoyannoj", Teor. Funkciĭ Funkcional. Anal. i Priložen. 17, 1973, 183-187.

A. Warzecha

[1] "The general form of subalgebras with unity of finite codimension of B-algebras with unity", Bull. Acad. Polon. Sciences, Ser. sciences math., astr., phys., 17, (4), 1969, 237-242.

J. Wermer

[1] Banach algebras and several complex variables, Markham, Chicago, 1971.

W. Żelazko

[1] Banach algebras, Elsevier Pub. Comp., Polish Sc. Pub., Warsaw 1973.

Notation

In this paper we shall use the following notation.

$\mathbb{N}$ = positive integers
$\mathbb{Z}$ = integers
$\mathbb{Q}$ = rational numbers
$\mathbb{R}$ = real numbers
$\mathbb{R}^+$ = positive real numbers
$\mathbb{C}$ = complex numbers

For a commutative Banach algebra A we denote:
$\mathfrak{M}(A)$ = maximal ideal space of A
∂A = Shilov boundary of A
ChA = Choquet boundary of A
A^{-1} = set of all invertible elements of A

For a compact Hausdorff space S we put:
C(S) $(C_R(S))$ = Banach space of all complex (real) valued continuous functions defined on S provided with the usual sup norm.

By a function algebra (= uniform algebra) A we mean any Banach algebra with unit such that $\|f^2\| = \|f\cdot f\|$ for any f in A. Any such algebra is usually identified, via the Gelfand transformation, with a closed subalgebra of $C(\mathfrak{M}(A))$ or $C(\partial A)$.

By $f|_K$ we denote restriction of the map f to a subset K of its domain.

For a Banach space X, δ_X denotes the modulus of convexity of X i.e. a function $\delta_X: \mathbb{R}^+ \to \mathbb{R}^+$ $\mathbb{R}^+$ defined by

$$\delta_X(\varepsilon) = 1 - \sup\{\tfrac{1}{2}\|x + x'\| : x, x' \in X, \|x\| = 1 = \|x'\|, \|x - x'\| \geq \varepsilon\}$$

we also define $\delta_X^*: \mathbb{R}^+ \to \mathbb{R}^+$ by

$$\delta_X^*(\delta) = \sup\{\varepsilon \in \mathbb{R}^+ : \delta_X(\varepsilon) \leq \delta\}.$$

Only notations which are used in more than one chapter are enumerated here. Thus, if you cannot find a symbol or notation in this table, look for it in the chapter you are reading. Note also that the index is an index of notation, too.

Index